SHROPSHIRE CANALS

AF735423

articles from the Shropshire Magazine

1950–1965

SHROPSHIRE LIBRARIES
1980

The publishers gratefully acknowledge the Shropshire Magazine's permission to reprint these articles, and the co-operation they have received from the Magazine's editor, Doreen Bourne, and printers. Thanks are also due to Russell Mulford for his foreword to the book, and to Sundorne Print, Shrewsbury for their part in the preparation of copy.

(Please note that any inaccuracies and uneven printing are in the original magazine articles. The publishers have not attempted to rectify these minor blemishes in producing this reprint).

Cover photograph: Tyrley Lock near Market Drayton
(by courtesy of the Planning Department of Shropshire County Council)

ISBN: 0 903802 11 2

Printed by Livesey Ltd., St. John's Hill, Shrewsbury

Published by Shropshire Libraries, London Road, Shrewsbury © 1980

Foreword

by Russell Mulford

For more than 100 years—from the early 1700's until the coming of the railways—the canal system of Great Britain provided the most important and economical method of inland transport in the country. Bulk loads which could not be carried on the roads of the day were easily transported, some of them with the aid of 'inclined planes'—a cleverly devised system of lowering goods from one level to another by means of rail-borne containers. Today the heyday of the canals is no more than a well-chronicled memory. Only a tiny percentage of the country's freight movement is now water-borne, although it must be said that there remain those who feel that the major canals may yet again come into their own. Fortunately many of the man-made waterways remain for pleasure use and many of the outstanding examples of canal architecture are there to be admired—like the aqueducts at Pontcysyllte and Chirk and on a different scale the iron aqueduct at Longdon on Tern, as well as the many bridges conceived by that gifted engineer Thomas Telford. Other canals, notably those in East Shropshire—the restored stretch near the new Coalport Museum apart—have fallen into disuse after being officially 'de-watered' by the British Waterways Board, though not without vociferous protests in many areas.

Generally, Shropshire had a most interesting network of waterways—notably those centred on Ellesmere and Whitchurch, the long since abandoned Shrewsbury branch, and the small network created to serve the industrial eastern part of the county. Each of these canals had its admirers if for different reasons and each has been placed under the spotlight in a series of articles which appeared in the Shropshire Magazine in the 1950's and 1960's.

E. A. Wilson was an authority on the Ellesmere network and his articles, reproduced in this booklet, reveal an intimate knowledge as well as a love of waterways. The same applies to W. Howard Williams, whose particular lifetime interest was the system built up around Donnington, Wombridge and Trench to serve local mines and industry of the Lilleshall Company and others. In contrast we have the writings of Lilian Hayward, who looked at the canals not as an industrial archaeologist but as the keen naturalist and walker that she was. Mrs. Hayward wrote her own obituary for the Shrewsbury Chronicle, to which she was a regular contributor, "in the interests of accuracy." Her writings on canal walks, in her own particular style, can here be savoured again.

Contents

Over The Shropshire Union Canal

STRONGLY reminiscent of his greater masterpiece at Pontcysyllte. this Telford bridge over the Shropshire Union Canal in Woodseaves Cutting, Market Drayton, shows the designer's positive genius with brickwork. Too often our early industrial buildings are written off as eyesores, but if one considers such works as this—or Brunel's famous "flat arch" brick bridge at Maidenhead—one must admit their exact proportions and undoubted beauty.

It is curious that the similar skill of Roman designers, as at the *Pont du Garde* in Provence, arouses immediate admiration; perhaps our Shropshire equivalents will be equally popular after another thousand years! In the meantime it is devoutly to be hoped that steps will be taken to ensure the preservation of Telford's work. (*Photo., Turner*).

Along Telford's Canal Path from Shrewsbury to Berwick Wharf

Seen from the Towpath........1 by LILIAN HAYWARD

I FANCY I hear someone say: "What! Go walking beside canals? Just straight ditches and nothing to see. How dull!"

Not so. The Shropshire Union Canals are by no means always straight. Sometimes they wind so much that they are more like rivers.

As they have been constructed upon high ground, they generally command fine views. Walking beside them we pass through charming, often beautiful, scenery. The quaint bridges, the little red houses which were built for the canal workers, and the occasional waterside villages give quite a continental effect.

Botanists will find an interesting flora of water plants and unusual birds may be seen. The tow-paths are easy walking and no bulls — the bane of walking in meadow-land — are to be met upon them.

For most of these rambles, which I shall describe in this and subsequent issues of the Shropshire Magazine, I had the pleasant companionship of my good friend Norah, who was kind enough to do some of the donkey-work, such as making the list of water plants, and going off to telephone about transport for the return journey, when we found ourselves finishing at unexpected places. She also carried the Thermos flask.

It is just as well to begin at Shrewsbury with the Shrewsbury Canal. This, with the bridges, tunnel and aqueducts, was constructed by the great engineer, Thomas Telford, between 1793 and 1797; it is a fine piece of work. The chief reason for its construction was to carry coal to Shrewsbury, mainly from the Marquis of Stafford's pits at Donnington Wood, for the roads were then inadequate for such heavy transport.

Donnington Wood is high above the start of the Shrewsbury Canal at Trench. In our third walk we shall see something of the ingenious method used to bring the coal from the high level down to Trench. Shrewsbury, being the objective, is really the end, rather than the beginning, of this canal.

We started from the canal bridge, opposite the Comet at Ditherington. It was formerly an exceptionally delightful walk beside the canal to Uffiington, returning by a ferry across the Severn. But now there is no longer a ferry boat and the water has been drained from the canal as far as Uffington.

However, it is still a pleasant walk, after a few minutes, getting away from the near vicinity of houses. The tow-path is on the right, or S.E., side of the canal. The Severn is beside the canal as far as Uffington, sometimes close to the canal, sometimes a field away.

From any part of the bank, above us, we had grand views of Pontesford Hill, the Longmynd, Wenlock Edge and Brown Clee, and especially of the Wrekin, with which we were to keep company for almost the whole length of the Shrewsbury Canal. Wooded Haughmond Hill was ahead.

Passing Pimley Manor, close to the canal, and Pimley Wharf, another ten minutes brought us to Uffington, after which there is water in the canal. (If you are out for a mere stroll you can walk up the road to Sundorne School, where there is an occasional bus back to Shrewsbury. From Sundorne Avenue, half a mile further, there is a frequent service).

Uffington is a long straggling village. There is a nice old farm-house set amid lilacs and laburnums, and a hospitable hotel, the Corbet Arms. The church was rebuilt in the 19th century. There are some charming German roundels in several of the windows and a case of curiosities in the N.E. corner.

For the next mile or so the scenery is delightful. Across the canal there is a near view of Haughmond Hill, with its woods and rocks. On this sunny day in Spring flag irises gleamed like gold above the blue water. The pretty little pink-flowered persicaria mingled with duckweed. Delicate sprays of water-plantain, and rose-coloured flowering rush were coming into bloom. Norah was very busy writing up her catalogue.

We passed under a number of bridges. In some cases the inside of the arch shows deep grooves, caused in the past by the chains or ropes of the canal boats. One of the pretty, old hump-backed bridges has been done away with, and the water is merely piped through it. One or two of the nice old lanes have been converted into motor roads. I believe this was done for the convenience of visitors to the Royal Show.

A goat was munching grass in the middle of the tow-path. We hoped that it was not a billy who would butt us into the canal. It turned out to be a tethered nanny, who graciously accepted a piece of cake.

We passed under a railway bridge into a deep cutting, and to the mouth of the tunnel, 970 yards long. There is no footway through the tunnel. The horses used to be led along a path overhead; the men in the boats worked their way through by pushing against the roof of the tunnel.

The overhead path has now disappeared and a strip of woodland grows above the tunnel. It is rather difficult to walk the whole way along the fields to the other end of the tunnel.

Norah and I compromised by cutting out the worst part and walking down a lane to Preston, a compact hamlet of red farms, one of them bearing a date stone 1856, and the initials R.B. (for Robert Burton of Longner) with the Burton coat-of-arms on a shield. We took another lane, left, back to the line of the tunnel and eventually reached the other end.

It is worth while, at a break in the strip of woodland, to cross it, to look at two large ventilation shafts, fenced round. Hart's tongue ferns grow inside the walls, and we could hear the water running far below. The arch at the far end of the tunnel is of red sandstone and bears the date 1797. The construction of this tunnel, over half a mile long, must have been a strenuous piece of work.

We walked through another deep cutting. There were masses of Herb Robert, and Norah found some plants with white flowers instead of the usual red. Comfrey was growing on the opposite bank. Coming up onto the road, when we returned to the tow-path, we found the remains of Berwick Wharf, where in the past the coal boats must have discharged some of their cargo.

This is not a pleasant part of the canal to-day. It is the receptacle for old tins and other refuse, thrown out from the houses which overhang the water. Great chestnut trees do their best to hide this foulness, their boughs laden with fragrant white blossoms, swept over the water.

In a few minutes we were away from this and were once more alone with flowers, butterflies and birds. A pair of swans, with six charming cygnets, sailed along. Papa hissed gently at us.

The next bridge, shored up with heavy timbers, marked the end of the ramble. Joining the road by a path on the far side of this bridge, we turned left and followed the road beside the high wall of Attingham Park, about a mile down to Atcham, whence there is an hourly bus service to Shrewsbury.

On the right we could see the Park of Longner Hall, with fine timber, which, however, hides the Hall itself. We passed Berwick grove and the gate of the drive — two miles long — to Longner Hall. On the left is the wall of Attingham and a very picturesque house, the Home Farm (with the delightful old name of Berwick Maviston).

Attingham, formerly the seat of Lord Berwick, is now in the care of the National Trust and is being used as an adult college. The distance from the Comet to Berwick Wharf, or rather the bridge beyond it and on to Atcham, is about six miles.

The entrance to the canal tunnel near Preston-on-Severn

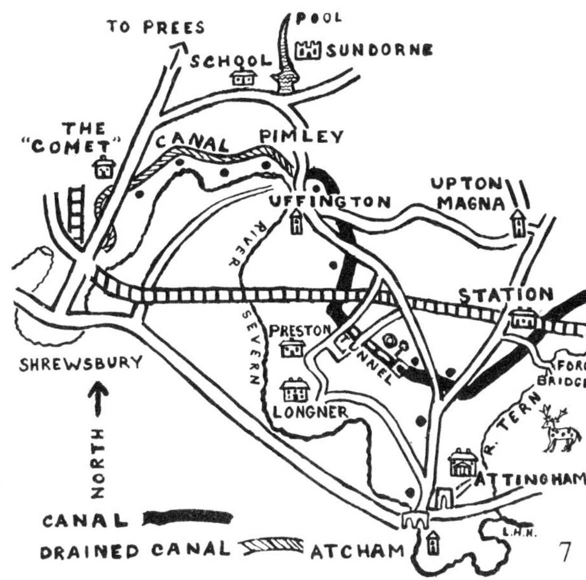

A Quaint Touch of Holland Between the Wrekin and the Ercall

by LILIAN HAYWARD

TO return to Berwick Wharf, or rather to the bridge beyond it, where we broke off our first walk, you can take the hourly Wellington 'bus service, get off at Atcham, and walk up the road past Attingham Park.

As an alternative, you can go by train to Upton Magna. Several trains stop at this station in the morning, but only one in the afternoon. Those who do not know Upton Magna may think it worth while to look round the village (half a mile away) and visit the Church. Not far away Forge Bridge, over the Tern, recalls the old forge where were forged the links of the suspension bridge over the Menai Straits, the famous bridge designed by Telford and largely cast by William Hazledine, the Shrewsbury ironfounder.

I will assume that you are back at the canal bridge where the first walk ended. All the bridges have numbers. This one is No. 37. The towpath is now on the right (S.E.) side of the canal. Norah and I started this walk on a beautiful June day. We met the same family of swans; the cygnets had grown a good deal. We saw several other swans, later in the day, but no more cygnets. We soon had a fine view of the conspicuous tower of Upton Magna Church.

We passed under a number of bridges, including one or two of the picturesque draw-bridges—tall wooden contraptions, made of heavy beams, worked by chains or wheels. They give quite a foreign effect.

Most of the road bridges are of brick, covered with a grey substance, so that in the distance they look like stone. Where farm tracks or footpaths cross the canal they are often made of woodwork, sometimes rustic work, but some of them have been replaced with concrete, not nearly so attractive.

Then there are iron bridges, some very plain, others of ornamental wrought-iron work. These fanciful bridges are generally to be found where the canal runs through some important estate.

At first Norah thought that it would be interesting to take a photograph of every different kind of bridge, but we found such a surprising number of types that the original idea would have been too heavy on films.

A pied fly-catcher showed itself. Butterflies and dragonflies flew along beside us. For about a quarter of a mile, a large shoal of fish swam along. A heap of broken snail shells showed where a thrush had been at work. Norah had hard work to keep her list of water plants. She normally walks much faster than myself, so she was able to linger over the water plants, and to catch me up, as I plodded along.

We met Mr. Grainger, of Withington, who looks after a four-mile beat. He keeps the canal clear, and cuts the grass from the towpath. It struck me that the upkeep of many miles of the now-unused Shropshire canals must be a considerable expense. Numbers of mussels had been dredged out of the canal along with the rushes, etc.

Haughmond Hill was now well behind us, and we were parallel with the Wrekin, its woods and rocks showing up well on this clear, sunny day. At times we seemed to have passed beyond the Wrekin, and then, as the canal wound north or south, we found ourselves opposite the Little Wrekin, only to be, apparently, opposite the Ercall a few minutes later. Not until the end of the day's walk did we feel that we had really put the Wrekin behind us.

Withington looks like a Dutch picture, the houses grouped around the red-sandstone spired church, seen across the canal. The church contains two interesting brasses: one (1515) to John and Joan Onley, with seven children, and a very strange elephant as an armorial bearing. The other brass is to Adam Grafton, who was incumbent here (1530). He was a great Pluralist, for, besides holding three livings, he was Master of Battlefield College, a Canon, a Prebendary, a Dean, an Archdeacon, and Chaplain to two of the royal Princes.

It is worth while to walk through Withington, which contains several pretty black-and-white houses, and to re-cross the canal and return to the towpath at the end of the village.

Another mile-and-a-half brought us to Rodington, on the River Roden. The church was built in 1851, and does not contain much of interest, but it stands in a good position, in the middle of the village, above a grass plot called the Bull Ring.

Near the Roden we met Mr. Parker, of Riverside House. He invited us to look at his very large garden, of 2½ acres, which he cultivates almost unaided, selling his produce in Shrewsbury. The garden was beautifully neat, everything growing well. It is one of those pleasant gardens where apple trees break the lines of the crops, and it is full of surprises.

Rodington is a small village but it has many amenities: two inns, a butcher, several drapers and grocers, a blacksmith and a wheelwright. Mr. Parker has a car for hire, and there is a 'bus service to and from Shrewsbury three times a week (the Green 'bus, which serves Withington also).

Tea can be had at the inns, but we had ours at the house of Mrs. Gourlay, Mr. Parker's daughter. The house is close to the canal, and was once an inn, catering for the people on the canal boats.

We left Rodington, much invigorated by an excellent tea. We crossed the Roden by a short aqueduct; here the canal and towpath are on the top of a bridge with three arches. We looked down on to the leaning fragments of what was once Rodington Hall.

The canal was running south-east, when it suddenly made a sharp bend and turned north-east. We saw various farms standing in the rich meadowland. Isombridge and Long Waste are quaint names. Sugden is an attractive-looking farm, with its group of buildings surrounded by trees. Large herds of cattle were grazing in the fields, or cooling themselves in the canal.

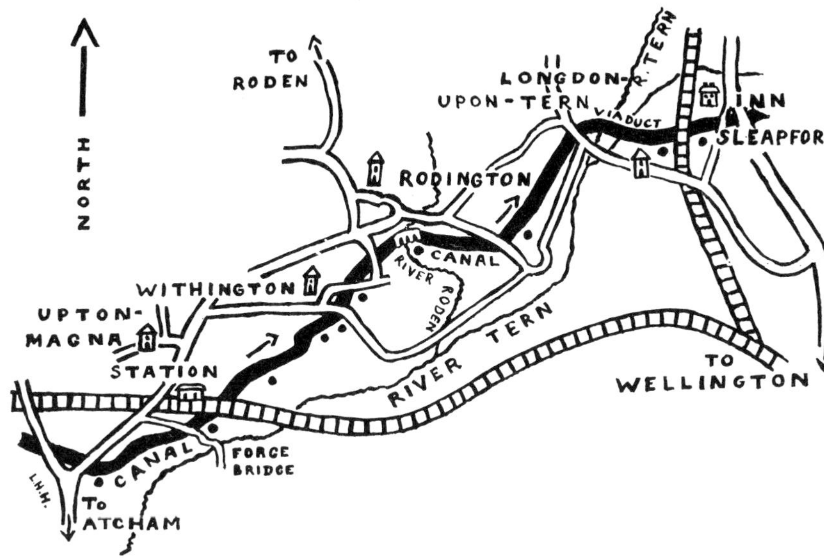

Soon we came to the aqueduct, 62 yards long, over the Tern. Here the canal runs through what might be described as a great iron trough, standing upon high iron supports, the towpath beside the water of the canal.

This aqueduct was designed by Telford and cast at Ketley, not far away, by Reynolds. Constructed in 1796, it was the first iron aqueduct ever made. Entirely of cast iron, except for the bolts, it is still in good working order. Longdon-upon-Tern, a small village near the canal, has a brick church, built in 1733.

Another mile brought us to Sleapford, where the bridge is inscribed "Long Lane, No. 16". So we had passed under 21 bridges, large and small, during the day's walk. Leaving the towpath, we reached the road near the Buck's Head, a long, attractive-looking inn, with a magnificent carved buck's head, with great antlers, as a sign over the entrance.

This is the stopping place for a not-very-frequent 'bus, direct to Shrewsbury, via Crudgington, Great Bolas and High Ercall (Services Nos. 922 and 923). In the opposite direction a 'bus runs more often to Wellington, where you must change. The distance of this walk is between eight and nine miles, including the piece of road at the beginning. It could, however, be divided into two walks by utilising the Green 'bus, which runs between Shrewsbury and Rodington, and in the reverse direction, on Wednesday, Thursday and Saturday.

A View from East of the Wrekin right to the Wall of Wales

Seen from the Towpath........3

by LILIAN HAYWARD

TO return to Sleapford from Shrewsbury, we took the 'bus labelled Great Bolas. After wandering through High Ercall, Crudgington, Waters Upton and Great Bolas (all pretty villages), this 'bus turned south for Wellington, passing through Sleapford, where we got out at the Buck's Head, to rejoin the canal at Long Lane Bridge (No. 16).

It was a beautiful day, clear and sunny. The Wrekin, which we had really passed at last, now well behind to the right, was bright green, with purple shadows.

The eastern ridge, leading up to the cone, looked quite jagged from where we now were. We could see the track up it quite plainly. Left of the Wrekin rose the wooded Ercall.

In the distance the Church Stretton Hills were pale blue, with deeper blue in the hollows. The cloud effects were beautiful. The water in the canal was bright blue, sometimes whipped up into waves by a strong South wind. Our way lay between cornfields.

We did not pass close to any villages, though Eyton-on-the-Weald Moors, with its little church, red houses, and Eyton Hall, were not far off. The eleven locks that we passed were the distinguishing feature of the day's ramble.

We passed two of the distinctive little red lock-keepers' houses. Presently we met Mr. Owen, sixty years of age, who has worked on the canal for 47 years, in fact since he was a boy of 13, as his father did before him.

He said that he thought it a pity that the canals have been allowed to go out of use for barge traffic. They could have been made use of during the last war, when transport was such a difficulty. But to-day many of the locks are quite dilapidated, and not fit to use. They are also so narrow that only the barges known as "narrow boats" could pass through them.

Mr. Owen gave us the names of the eleven locks, some of which are very quaint. These names are : Eyton Lower Lock, Eyton, Wappenshall, Britain, Wheat Leasowes, Shucks, Peaty, Hadley Park, Turnip, Castle, and Trench. He lives in one of the lock-keepers' little houses that we had passed.

We ate our sandwiches close to Eyton Lock. Throwing a crust into the water we were amused to see a swirling circle appear in the lock, and then a grey snout came up to annex the crust. Every bit that we threw into the water was thus seized.

We thought that the fish might be pike, but two anglers close by said that they would be roach. As, however, they admitted that there are pike in the canal, I still have my doubts. They were fishing for roach or perch, as well as for such small fry as dace.

The nearer we got to Trench the more fishermen we passed—and the younger. We did not see a fish landed, but one urchin had a good sized roach in his creel.

Two miles from Sleapford, we came to Wappenshall Junction, the junction of the canal along which we were walking with the branch to Newport. We looked along a mile stretch of blue water, with a pretty grey, arched bridge in the distance.

A not unpicturesque red building, apparently once a mill, stands above the junction of the waters. A very curious massive stone walled footway leads up to the level of the first floor of the mill, crosses the water, and comes down again to the tow path. We wondered if it had been really necessary to construct such an elaborate and massive affair.

This seemed the great time for water plants. Masses of hemp agrimony grew on the opposite bank, its pinky mauve contrasting with small yellow water lilies.

The rare Villarsia, which has disappeared from parts of the canal, was here in great quantity. This has flat roundish leaves, crinkled at the edges, and small yellow blooms, which might be mistaken for a kind of water lily, but the petals are more pointed.

Here, for the first time, we saw the arrowhead, with noticeable leaves of that shape, and the white flowers just coming into bloom. Blue skullcap, gipsywort and peppermint were

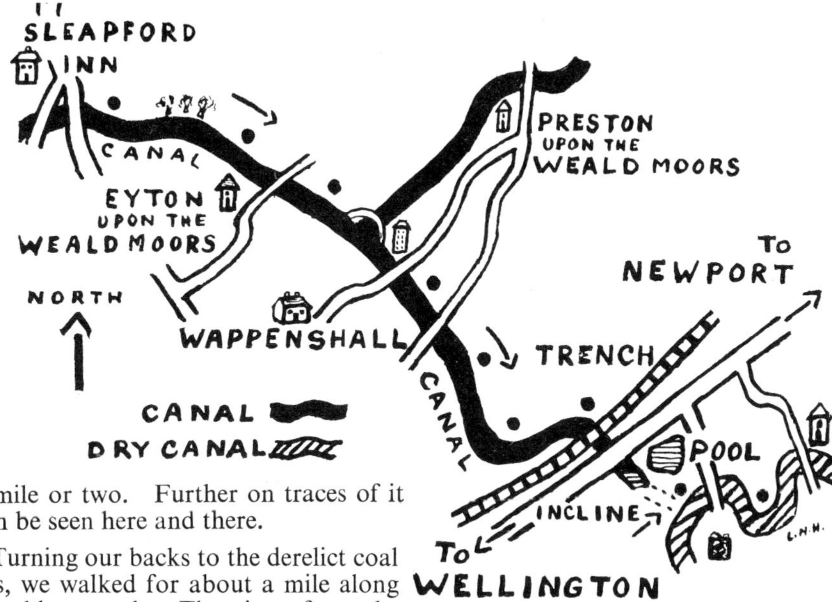

abundant on the edge of the water, and the leaves of the great water dock, some as much as two feet long, as well as many other water-loving plants.

The canal passes under the Shrewsbury-Newport road, and we were soon on a footpath above what was the beginning of the Shrewsbury canal, now choked up with rushes, etc.

The canal used to be fed from Trench Pool, on our left, a fine piece of water where boys were bathing. Coots were swimming in and out of the reeds, giving shrill cries to one another. Swallows were skimming over the water, and what we hoped were sedge warblers were twittering among the reed maces (bulrushes).

Passing a small inn, we took the footpath which led up beside the remains of the inclined plane which formerly connected the Shrewsbury canal with the Marquis of Stafford's canal, on a higher level.

This canal, or "cut," as it is locally called, the first to be made in Shropshire, was constructed in 1782. It was made to carry coal from his Lordship's pits to his lime works at Lilleshall, as well as to Pave Lane (near Newport,) where the wharf and basin can still be seen.

To convey coal to the Shrewsbury canal, it was loaded into tub boats at the top of the incline. These boats were taken down full, and brought up again empty by means of cables worked by an engine, the engine house being at the top of the incline.

When we reached the head of the incline we found ourselves beside the Marquis of Stafford's canal. It ran for seven miles on a dead level, as far as Pave Lane. Water was pumped into it from the coal pits. It is now dry, but there is a pleasant walk alongside it for a mile or two. Further on traces of it can be seen here and there.

Turning our backs to the derelict coal pits, we walked for about a mile along the old towpath. The views from the high level are magnificent, in the west stretching away to the wall of Wales, the Berwyn Mountains. We also had a fine and unusual view of the Wrekin, its cone rising above the Ercall like a real mountain.

Besides the coal, which was the chie reason for the construction of the Marquis of Stafford's canal, other goods were taken down to the inclined plane, flour from the mill at Wrockwardine Wood, and other commodities, the greater part of which were destined to go to Shrewsbury, and, later, to Newport.

It must be remembered that at the time when the canals were made there were no railways, and it was almost impossible to carry heavy goods, such as coal, upon such roads as then existed.

There soon followed several other canals, of which few traces remain to-day, between Trench and the Severn. These were made to facilitate trade, and especially in connection with the iron foundries, in the industrial area of Shropshire.

There is no barge traffic upon the Shrewsbury canal to-day. The only craft to be seen are the boats used by the canal keepers. Despite the efforts of these men the canal is, in places, choked up with sedges and rushes.

We walked back to the head of the incline down to Trench Pool, and on to Trench Crossing, from which point there is an hourly 'bus service back to Shrewsbury. The halt is just outside the "Barley Mow."

The distance from Sleapford to Trench is about four and a half miles. Norah and I put in another two or three miles, walking up to the High Level, alongside the Marquis of Stafford's canal, and back to Trench.

Where once the "Eys" rose from the marshes

by LILIAN HAYWARD

ON our third walk it will be remembered we passed by Wappenshall Junction, where the water runs swiftly below an old building. For our next walk we decided to return to this point and walk along the towpath to Newport.

Unless you have a car, or a kind friend who has one (as we had), you will have a mile or two to walk from the 'bus to Wappenshall Junction. The Great Bolas - Wellington 'bus comes nearest, but if you prefer to take the direct 'bus to Wellington, there is a pleasant walk, though longer, through Apley Park to the canal.

From Wappenshall Junction there is a straight mile to the next bridge, after which the canal winds near the pleasant village of Preston-upon-the-Weald Moors.

The chief point of interest here is the "Hospital" for elderly ladies and children. This was founded in 1716, under the will of Lady Catherine Herbert, daughter of the first Earl of Bradford, and widow of the last Lord Herbert, of Chirbury.

Her brother, Lord Torrington, (second son of Lord Bradford) also gave £1,000 for building the dining hall, in which portraits of himself and his wife are hung.

Just beyond Preston Bridge, we came to a very pretty reach of the canal, which here widens out below trees, giving the effect of the approach to a Thames back-water. In less than a mile, on the opposite side, a feeder to the canal runs in from near the Humbers. A farm with a quaint name, Lubstree Park, is not far away.

Soon the canal passes as a very short aqueduct along the top of a stone bridge above a roadway. The ancient arms of Sutherland are carved upon the arch of the bridge.

A canal keeper whom we met told us that the road below was formerly a private road from Lilleshall, made by "the old Squire" (by whom we supposed that he meant one of the Dukes of Sutherland).

From this bridge we had a good view of the Wrekin behind us, and on the right a ridge broken up into small wooded heights. This high ground seems to divide the Weald Moors from the Severn Valley.

We were now crossing an interesting tract of country, the Weald Moors, or Wild Moors—that is waste, or uncultivated land. This district is contained, roughly speaking, within a triangle of which the points are Newport, Wellington and Crudgington.

Much of it lies low, and before it was drained was marshy and wet, with only a few settlements, rising as " eys," or islands, of higher ground, such as Kynnersley (a mile from the canal), Adeney, Butterey and Eyton.

A rector of 200 years ago described Kynnersley as being always surrounded by water in winter. The Wild Moors were then overgrown with scrub, willows and brushwood.

The cattle turned out to graze had bells hung to their necks, so that they could be found at night. Bullocks did particularly well on this rough grazing.

Half-a-mile beyond Kynnersley is Wall Camp, with extensive earthworks. A glance at the O.S. map shows this region criss-crossed by dykes and strines. The draining of the marsh land by the first Duke of Sutherland, in the early part of the last century, entirely altered the character of the Weald Moors, and brought hundreds of acres into cultivation.

Near Buttery Farm we crossed the canal by a curious bridge, divided by a wall into two passage ways, a narrow one for the canal boat horses (for the tow path now passes to the other side of the canal), and a wider one for wheeled traffic.

We passed Adeney, and hereabouts we saw great masses of silky headed reeds, nearly filling up the canal from bank to bank. When we came back to open water there were many swans swimming up and down. Indeed what in the distance looked like linen spread out to dry on the bank, turned out to be a group of swans lying close together in the sunshine.

After passing under one or two narrow bridges, we came to a wider one over which runs the road to Edgmond, less than a mile away. Just beside the bridge stands the very picturesque Bridge Farm.

Those who do not know the fine old church of Edgmond should on no account miss seeing it. It contains a very remarkable carved font, thought to be nearly a thousand years old, as well as what is perhaps the finest brass in the county, dated 1533. This depicts Francis Yonge (shrouded), his wife Margaret (daughter of Richard Eyton)

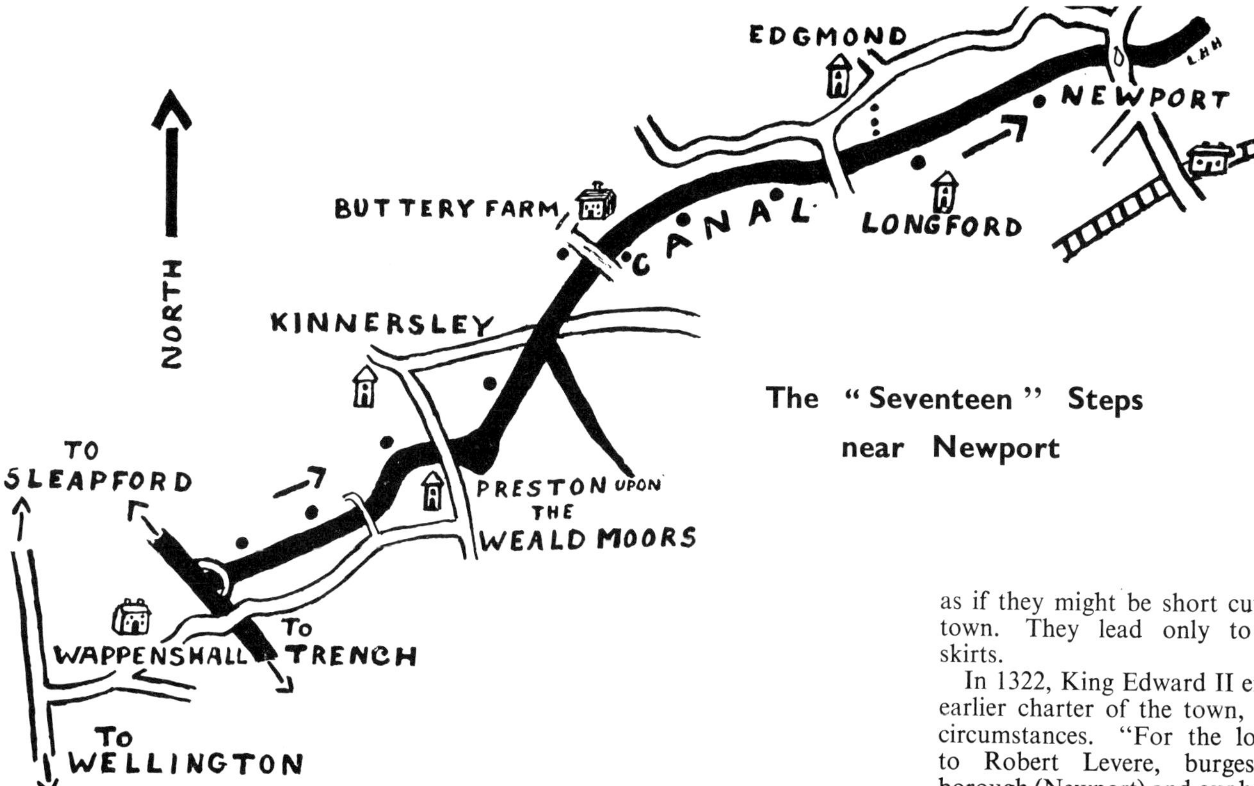

The "Seventeen" Steps near Newport

with their sons and daughters.

You can return to the canal by the same road; by the footpath (if you can find it); or you can shorten the walk by taking the 'bus into Newport. This is the same 'bus which goes on to Shrewsbury. The Suffragan Bishop of Shrewsbury resides at Edgmond.

Norah and I had both been to Edgmond before, so we kept to the tow path. We had a view of the park and part of the house of Longford Manor. The church was rebuilt 150 years ago, but some remains of the earlier church can be seen in the churchyard.

We also had a good view of Lilleshall Hill, surmounted by an obelisk, erected in memory of the first Duke of Sutherland by his tenants, according to the custom of the period.

We passed three locks, the middle one known as Polly's Lock. I wondered who Polly might have been—perhaps a former lock keeper. Soon we were encouraged by the sight of Newport Church, its fine tower looking much nearer than in fact it was. We passed a very pretty little red house with a gay garden—perhaps one of the canal keepers' dwellings.

We left the canal at the point where it is crossed by a main road. By the way, do not be tempted to take any of the footpaths on the right, which look as if they might be short cuts into the town. They lead only to the outskirts.

In 1322, King Edward II enlarged an earlier charter of the town, in curious circumstances. "For the love I bear to Robert Levere, burgess of the borough (Newport) and our host there"; when, in the fifteenth year of his reign the king "lay at the Antelope in Newport, one Robert Levere, a merry host being master of the inn, who so pleased his majestie in his entertainment that for the sake of this jolly landlord, the town had their charter renewed."

The arms of Newport are three fishes, which are thought to refer to the famous vivary (or fishpond), adjoining the town from which the burgesses had to supply the King's Court with fish.

We hastened up the hill into the town, past the seventeenth century Grammar School, founded by William Adams, a member of the Haberdashers' Company, past the church, to the stopping place of the hourly 'bus to Shrewsbury, outside Brittain's Stores, a few minutes before it was due.

It was now that Norah displayed great resource. We were longing for tea, but could not remain another hour in Newport to get tea there, and return by a later bus. Nothing daunted, Norah hurried along the street, discovered a cafe-canteen, ordered two cups of tea, which she poured into the thermos flask, seized two large buns, and arrived back just as the 'bus drew up. What if the tea did taste slightly of the coffee which had previously been in the flask? We were very thirsty, and seldom has tea been more acceptable.

The distance from Wappenshall Junction to Newport is six-and-a-half miles. There is an interesting walk of four miles on from Newport to Norbury Junction, on the main canal.

There are twenty-one locks in this four miles. Of these twenty-one locks, seventeen of them come close together to give a substantial rise to higher country. They are locally known as "The Seventeen Steps."

I have been unable to find out whether there is a 'bus service on the Newport-Eccleshall road, about a mile from Norbury.

In any case, the best walk would be from Newport to Norbury, and then along the main canal southwards to Gnosall, from which station the return journey can be made by train.

Much of the tow path from Norbury is through woods, and there is still a certain amount of barge traffic on the main canal. Here, with luck, we may see some of the gay canal boats, painted with roses and castles in the traditional manner.

Drawbridge over the Canal near Whixall Moss

Above Whixall Moss—haunt of rare Marsh Plants

by LILIAN HAYWARD

I NOW turned north and west, alongside what is known as the Ellesmere Canal, though still a part of the Shropshire Union Canal, which eventually took me into very beautiful Denbighshire scenery.

This canal begins a few miles beyond Llangollen, taking water from the River

Dee, and, passing near Ruabon, Chirk, Ellesmere, and Whitchurch, joins the main line canal at Hurleston, not far from Nantwich.

From Nantwich, the course of the main canal is northwards to Chester and Ellesmere Port, near Birkenhead, and southwards by Market Drayton and Norbury Junction to Wolverhampton and Birmingham, with connections to the Staffordshire and Worcestershire, the Grand Trunk, and other canals.

There is very little traffic upon the Shrewsbury and Ellesmere canals, but there is a considerable amount on the main line canal.

Telford was responsible for practically the whole of the Shropshire Union Canals. His construction of the Ellesmere Canal, with its embankments, tunnels and aqueducts, between 1793 and about 1805, was one of his greatest works.

To reach Whitchurch from Shrewsbury there is the choice of train or Midland 'Bus. The 'bus is slightly more convenient as regards time service; also its terminus in Whitchurch is much nearer the canal than the railway station.

But, as the return journey must be, at least partly, by train, it is more economical to travel both ways by train, with a day ticket return, Shrewsbury to Whitchurch, and at the end of the walk with a single from Bettisfield to Whitchurch.

The walk from Whitchurch to Bettisfield is about ten miles. The distance can be shortened by taking the train on from Whitchurch to Fenn's Bank station. Starting from there, you will have the best part of the walk. Fenn's

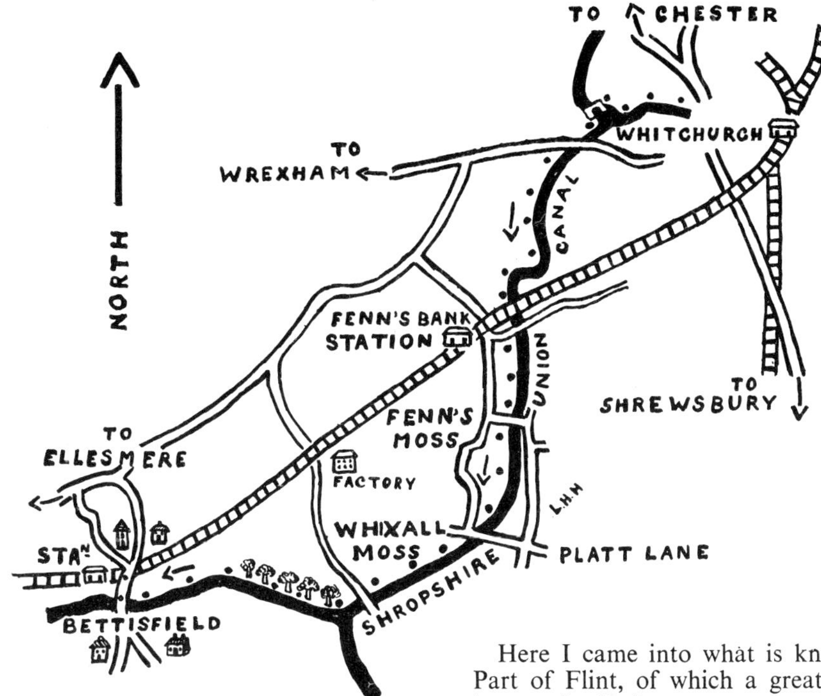

Bank and Bettisfield stations are both on the Whitchurch-Oswestry line, on which there is a fairly good service.

My kind comrade could not accompany me on this walk, and I missed her company, as well as her botanical knowledge. From Whitchurch I started along the short canal branch—about a mile—to its junction with the Ellesmere Canal.

North-east, the Ellesmere Canal passes through flattish country to join the main line canal some sixteen miles away. My own way was in the opposite direction, south-west; first through well timbered meadow-land, the canal twisting about, under humped-back bridges, until near Fenn's Bank station.

Here I came into what is known as Part of Flint, of which a great wedge inserts itself into our county. Not until after Bettisfield does the canal come back into Shropshire.

The tow path is on the right, or north-west bank of the canal. The Ellesmere Canal is wider than the Shrewsbury Canal.

Near Fenn's Bank I was delighted to come upon my first barge—the Pollux—belonging to British Railways, which now own the canals. The Pollux is one of the craft kept in use to keep the canal clear, and to repair the banks, and was loaded with clay for this purpose.

Although the canal is now little used, its water, coming from the Dee, among other sources, helps to feed the main line canal. Though rather dull in appearance, the barge being painted black, the tiller, etc., were gay with blue and yellow.

There was a cabin on board, where the crew of two slept and cooked their meals, returning to their homes at the week-end. The horse which draws the Pollux was grazing peacefully beside the tow path.

I was now walking beside a pretty, river-like stretch of water. By this time the water plants had nearly finished, although a few blooms of monkey-plant and gypsywort were still to be seen. The "snowballs" of the wild guelder rose were already replaced by clusters of red and yellow berries.

Swifts and swallows were scudding over the water, and I was thrilled to see at least four sedge warblers—gay little birds, whose reddish-brown backs were brilliant in the sunshine.

At "Brick Walls" I passed one of the picturesque wooden drawbridges, and beneath several stone ones. From the crown of one of these I had a fine view of the Berwyns ahead, while the Breidden summits appeared rather unexpectedly, away on the left.

From Platt Lane, about half-way between Fenn's Bank and Bettisfield, the canal skirts Whixall Moss, the haunt of the rare marsh plants. Just here, the canal is constructed upon the top of a high embankment.

From the tow path I looked down upon a number of rose-covered cottages, each with its little stack of "turfs." Peat is dug from the moss, and dealt with commercially in a factory on the far side.

If we should wish to cross to the Wrexham road, we should take the track opposite Moss Farm, and go by way of Oaf's Orchard and Cuckoo's Corner. But to-day I kept to the tow path.

I passed another great wooden draw-bridge, and the junction with a branch, or feeder, of the canal, ending (or perhaps I should more correctly say beginning) at Quina Brook, near Pepper Street. There are centainly some quaint names in this district.

Now I had a mile long stretch, bordered by trees which gave pleasant shade upon this hot August day. Several swans were on guard, their families no doubt hidden among the reeds and rushes on the opposite side of the canal. One honked at me, and another hissed at me, but they did not otherwise interfere.

Another canal boat, long and narrow, the Venus, lay right across the canal, reaching from bank to bank. There was nobody on board.

I was nearing Bettisfield. On the right I could see Cornhill Farm, and on the high ground above the farm the turret spire of the church, and the railway station.

I noticed a footpath which cut off a corner, and made a short cut to the station. But I also noticed that in the same field there was a large herd of cows, and a bull with them, so I thought it wiser to keep to the tow path until I reached the road.

This was further than I had expected, and I had to hurry up the hill to catch the train for Whitchurch. On the right of the road was the field with the footpath, cows and bull.

Now, would you believe it? When I was halfway up the hill, lo! the bull walked through a gap in the hedge onto the road, just in front of me, and began grazing on the verge.

What was I to do? There seemed no other way. The only possible course was to walk past him, which I did, keeping close to the opposite hedge, where there were some rails, which I felt that I could climb quickly, if necessary.

He was quiet enough, however, poor brute, being entirely occupied with the bit of fresh grass which he had discovered—a refreshing change from his burned up pasture.

I just caught my train for Whitchurch, where I changed for Shrewsbury. It was interesting to look out of the carriage window, and to get an occasional peep at the canal, beside which I had been walking.

Seen from the Towpath......6

Pretty walk where the Ellesmere canal winds among the Meres

by LILIAN HAYWARD

THE natural thing would have been to continue the walk from Bettisfield to Ellesmere, but the journey by 'bus and train is so much more convenient taken in the reverse direction.

A 'bus leaves Shrewsbury for Ellesmere at 9.10 on Monday and Saturday, or at mid-day on Tuesday. (As always please verify these times, which may alter from time to time). It is a pleasant run through several pretty villages, Albrighton, Preston Gubbalds, Myddle and Cockshutt.

After Cockshutt, there is a glimpse of Crosemere on the right, and, about two miles further, Whitemere can be seen on the left. At the next main road fork, about three-quarters of a mile short of Ellesmere, we joined the canal.

Those who do not already know Ellesmere, beautifully situated beside and above the mere, may like to go on to the 'bus stop. Across the mere, on the opposite bank, is Oteley Park, with the fine house built rather more than a hundred years ago.

A White Lady was said to haunt the former house; so much so that when the present mansion was built, a piece of the old house was left standing, in the hope that this would encourage the White Lady to stay there, and not move on to haunt the new home !

Ellesmere, together with the Castle, of which nothing remains to-day except the mound on which it stood, was, in 1205, settled upon Llewellyn, Prince of North Wales, when he married Joan, a daughter of King John.

The church stands on high ground, not far from the site of the Castle. There is a lovely view of the mere from the churchyard. Although the church was considerably rebuilt during the last century, some interesting old work remains, and especially the Chapel of St. Anne (generally known as the Oteley Chapel) which has an exceptionally fine roof with carved beams, about four hundred years old.

In this chapel is an altar tomb wth the carved figures of Sir Francis Kynnersley, who was cup bearer to Queen Elizabeth, and his wife Margaret, daughter to Sir Francis Charlton. Sir Francis is represented in Elizabethan armour, with ruffs. Lady Margaret wears a Mary Queen of Scots cap and neck ruff.

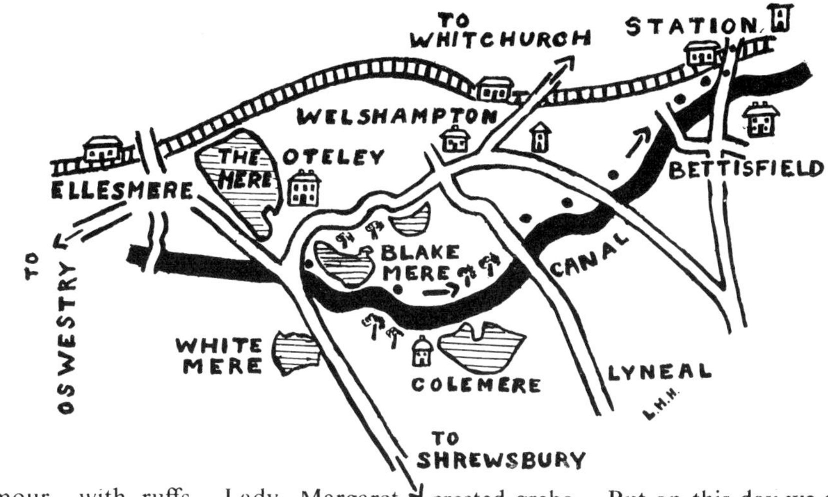

On the south of the Chancel it may need rather a hunt to find the quaint diminutive figure of a fourteenth centruy notary (or lawyer) holding a book; his inkhorn and penner hang from his girdle. This little figure is almost unique.

Having seen Ellesmere, return to the fork road, and descend to the canal, here in a deep cutting. On the left, or north, of the canal, it may be interesting to look back along a tunnel, or even to walk through it, as there is a footway. But we shall find better or, at least longer, tunnels another day.

For the next mile or so, the stretch is one of the prettiest on the canal. Clear water, moving gently along, is overhung by trees. There are quantities of ferns growing on the banks.

We came upon another barge, with another astronomical name, the Arcturus. The canal runs alongside Blake Mere, a charming sheet of water, with a background of trees, where one may sometimes see pochards and great crested grebe. But on this day we saw only swans. Perhaps the rarer water fowl had retreated from the laughter of a gay party out in a row boat.

A fleet of eighteen ducks were swimming on the canal, keeping pace with us and, perhaps, hoping for some tit bits. At length they left the water to trail through the grass to a charming thatched house on the other side of the canal.

The owner of this house must have a delightful view of Colemere, of which we could catch glimpses on the right through the trees.

Walking beside Colemere on a former occasion, a number of great crested grebe were both diving in the mere, and flying across it.

Beyond the meres, the canal winds a good deal, sometimes with quite sharp bends. A strong wind was blowing. This swept along the now unsheltered tow path.

The bridges are arched structures, of brick, covered with a grey substance, like some of the bridges on the Shrewsbury canal. Several of them have curious alcoves to hold the boards which are—or were—used at times to dam back the water.

In other places we had seen these boards stacked on the side of the canal with a sort of pent roof over them, the whole contraption having rather the appearance of an old-fashioned goose cub.

Among the flowers, Norah found the not so common fleabane, restharrow, and burr marigold. A great number of dragon flies accompanied us, as well as pretty little "gate keeper" and skipper butterflies.

Presently we came upon a lorry containing a tank, and a number of "churns," which a farmer and his man were laboriously filling with buckets of water from the canal; the pond of his farm having gone dry.

We left the canal at the road bridge just above the house marked upon the map as "Inn" and, once again in part of Flint, we walked up to Bettisfield Station.

As we had a little time to spare we went to see the church, close by. On the way we passed Bettisfield Old Hall, an attractive old brick house, with mullioned windows. A newer wing has been added. Unfortunately, the old part of the house is in a poor state of repair.

The church was built by Lord and Lady Hanmer in 1880. For a comparatively new building it is very pleasing, built of warm pinkish-grey stone. It has a small round tower. Inside all is in good keeping, simple and in good taste, and especially the pulpit. There is a view of distant hills from the churchyard.

There is no 'bus service anywhere near Bettisfield, indeed, there is not even one between Oswestry and Whitchurch, so we returned by train to Shrewsbury, changing at Whitchurch, where there was time to have tea. It is only between five and six miles from the fork road, near Ellesmere to Bettisfield, but the energetic walker might combine this with the last walk, or at least as far as Fenn's Bank.

Seen from the Towpath......7

Where canal boats float high above the rivers

By LILLIAN HAYWARD

IT was not possible to walk every mile of the tow paths beside the Shropshire Union Canals, so we gave a miss to the district between Ellesmere and Gobowen, which is rather flat country, though even this has its own charm.

The distance between the Moors, near Gobowen, and Trevor is about seven-and-a-half miles, or nine from Gobowen. From Gobowen the Crosville 'Bus Company runs a service to St. Martin's each afternoon at 2.9, and on Saturdays at 11.54, with a halt at the Moors. These times should be checked in case of changes.

There are good train services, both ways, between Shrewsbury and Gobowen, and fairly good train and 'bus services back to Gobowen from Chirk and Ruabon.

Nora and I had another companion on this walk—a botanical friend, Ellen, who knows the district well. We left the 'bus at the Moors, just by the old hump-backed canal bridge.

It is remarkable that these old bridges, which were intended to carry nothing heavier than a wagon or coach, now bear the weight of double-decker 'buses, travelling at high speed. The old masons built well and truly.

We passed "The Shop," to which, Ellen told us, goods were brought by the barges, and the near-by cottages were then the homes of the canal folk.

Our walk began in Shropshire and ended in Denbighshire. The canal wound along beneath trees, with views of hills to the south. One of the bridges under which we passed is crossed by Watt's Dyke, an ancient boundary, possibly made by the Saxon leader, Wada.

The raised bank under the beeches is still known as the "Dyke Walk." The better known "Offa's Dyke" is only a few miles away.

On the opposite bank I noticed a pretty white gabled house, flanked by yews. As, later on, I saw a very similar house beyond Chirk, I thought that they might be the homes of men connected with the canals.

We passed the villages of Rhosweil and Gledrid, where roses and nasturtiums hang over garden walls, and stone steps lead down to the canal from which the cottagers draw water.

After Gledrid the canal is close to Telford's Holyhead Road. Note the Woodland's Cafe, where refreshments can be obtained.

Just where the "new" Holyhead Road dips down into the Ceiriog Valley, the canal, high above this valley, is crossed by the "old" Holyhead Road. Ellen said that extra horses used to be harnessed to draw the coaches up the steep hill.

The substantial house at the bridge

head was formerly a coaching inn. Across the valley we had a pleasant view of Bryn Kinalt, Lord Trevor's place, and the steeple of the church of Chirk.

Just short of the aqueduct over the River Ceiriog, we came upon quite a fleet of barges, belonging to the British Railways, including my old friend, "The Pollux," which I last saw near Fenn's Bank, "The Emerald," gaily painted yellow and blue, and "The Swift."

We crossed the aqueduct, and looked down on to the lovely river below. To the left are the much higher arches of the railway bridge which rather block the view up the valley.

Hundreds of swallows nest on the parapet of the railway bridge, and in the Spring are very busy, flying to and fro, to feed their young.

The Ceiriog is said to be the boundary between Shropshire and Denbighshire. In fact, it is the *old* bed of the river, which has altered its course, that is the boundary.

The tow path between the canal and the outside of the aqueduct is protected, not only by the iron railings, but also with stout wire netting; for it is told that once upon a time a small girl fell through the railings. It was the era of many starched petticoats. These floated the child down like a parachute, and bore the small wearer uninjured down to the valley, 65 feet below !

From the Denbighshire end of the aqueduct I had to finish the walk by myself. The canal now passes through a tunnel, about a quarter-of-a-mile long.

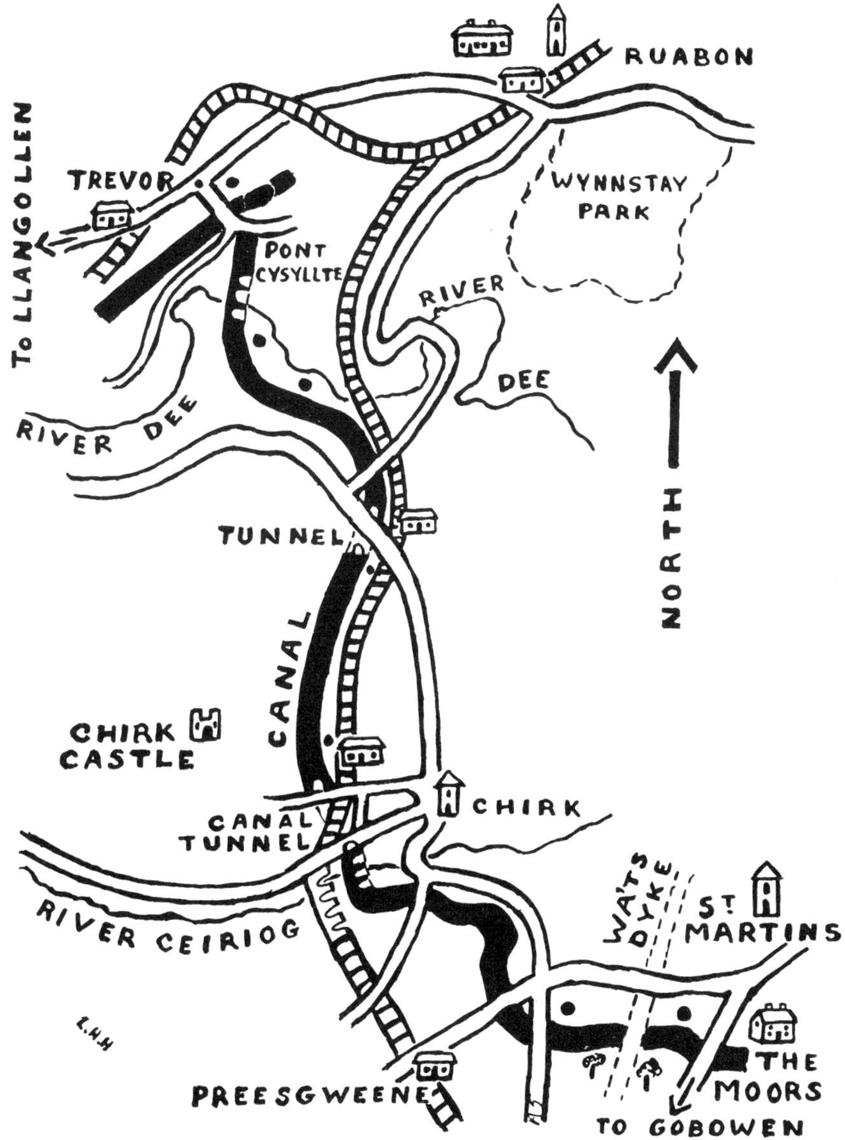

From the mouth of the tunnel the half circle of light at the other end showed clearly, so I started gaily along the raised footway, not in the least realising what a long way it would be to the other end. It was quite safe, with a strong hand-rail protecting the footway, along which the canal boat horses used to be led, from the canal.

Nevertheless, I found it rather an eerie experience, in the half light, with the dark water beside me, and more water dripping from the roof.

Those who do not fancy the idea of this tunnel can walk along the road overhead, and re-join the canal to the left of Chirk railway station.

I reached the far end of the tunnel at last. I was now in a deep cutting, where the canal, winding beneath trees, and edged with ferns, flowers and rocks, resembled a river.

Then, once more I was in the sunshine, between meadows, with Chirk Park on the left. Chirk Castle, home of the Myddleton family for nearly 400 years, cannot be seen from the tow path.

It was a hot day; family parties were lunching beside the canal, and boys were bathing. Here, the canal contains clear Dee water, constantly flowing, and renewing itself, unlike the stagnant ooze in parts of the Shrewsbury Canal.

I passed under perhaps a dozen bridges and through a short tunnel; after the last bridge I had a wonderful view of the Dee Valley, and the not unpicturesque railway viaduct crossing it on the right.

Soon after passing some quarry machinery there is a drawbridge on the left, leading to a basin. The way to Telford's famous aqueduct at Pont Cysyllte is straight on. I, however, crossed the drawbridge, because I had been told that the "Cressy" was lying in the basin.

The "Cressy" is the *Narrow Boat*, about which Mr. L. T. C. Rolt has written a particularly delightful book. It was, in fact, this book which prompted me to start walking alongside the canals.

I had expected to find the "Cressy" painted all over with roses and castles, but Mr. Rolt had used a good deal of restraint on the outside, although traditional patterns decorate the interior. The owner was on board, and I had a chat with him.

The "Cressy" was originally a "Shroppie Fly-boat"—that is built by the Shropshire Union Canal Company, for their "fly" service, which operated day and night, with relays of horses.

After several intermediate uses, the "Cressy" was bought and equipped by Mr. and Mrs. Rolt, who use her as their permanent home, and have explored many canals in her.

I returned to the tow path and soon reached the aqueduct—the second on this walk. It was one of Telford's great achievements. It is a quarter-of-a-mile long and, in the middle, is 150 feet above the Dee Valley. Like the Chirk Aqueduct it is well protected, and is quite safe. But I have no head for heights. I hated the feeling of being suspended in mid-air. I forced myself along only by clutching the railing.

A workman whom I met stopped and asked me, " Don't you like it, Ma"? " No," I replied, "I'm scared stiff."

He obligingly turned back, " to see me across." I could have wished that he had not been so anxious for me to admire the beauty of the Dee below, for to have looked down would have about finished me.

To others with weak heads, I may say that it is possible to avoid the aqueduct by going down into the valley, crossing the Dee by a road bridge, and climbing up again. This, however, is an arduous and circuitous route.

Once across the aqueduct I was in the village of Trevor. A few minutes walk uphill brought me to the Crosville 'bus halt, and to the railway station, from either of which Ruabon can be reached in about a quarter-of-an-hour.

Seen from the Towpath......8

Where the Ellesmere canal winds in glorious scenery

by LILIAN HAYWARD

FOR the last section of the Ellesmere Canal, I found it more convenient, to stay the night at Ruabon.

When I formerly passed through Ruabon in the train I had the impression of an industrial district beneath a pall of smoke. I was, therefore, agreeably surprised to find Ruabon—or at any rate the part that I saw—looking like any small market town, with a brook beneath a stone bridge, and a hill leading up to the fine church, a few shops, and the Georgian Posting Hotel, where I stayed.

Among the guests was an interesting party, a husband, wife, and the husband's mother, all upon a walking tour, including the mother, aged eighty, who could still do her seven or eight miles a day.

Their method was to walk three or four miles in the morning, lunch at a hotel where the old lady could rest during the afternoon. Then, after an early tea, they would walk on to the place where they had arranged to spend the night. If the distance worked out at more than eight miles, they would arrange for a car to meet them.

After breakfast I returned by bus to Trevor, to walk to Llangollen, and eventually to reach Berwyn, the headquarters of this canal. I was in Denbighshire practically all day,

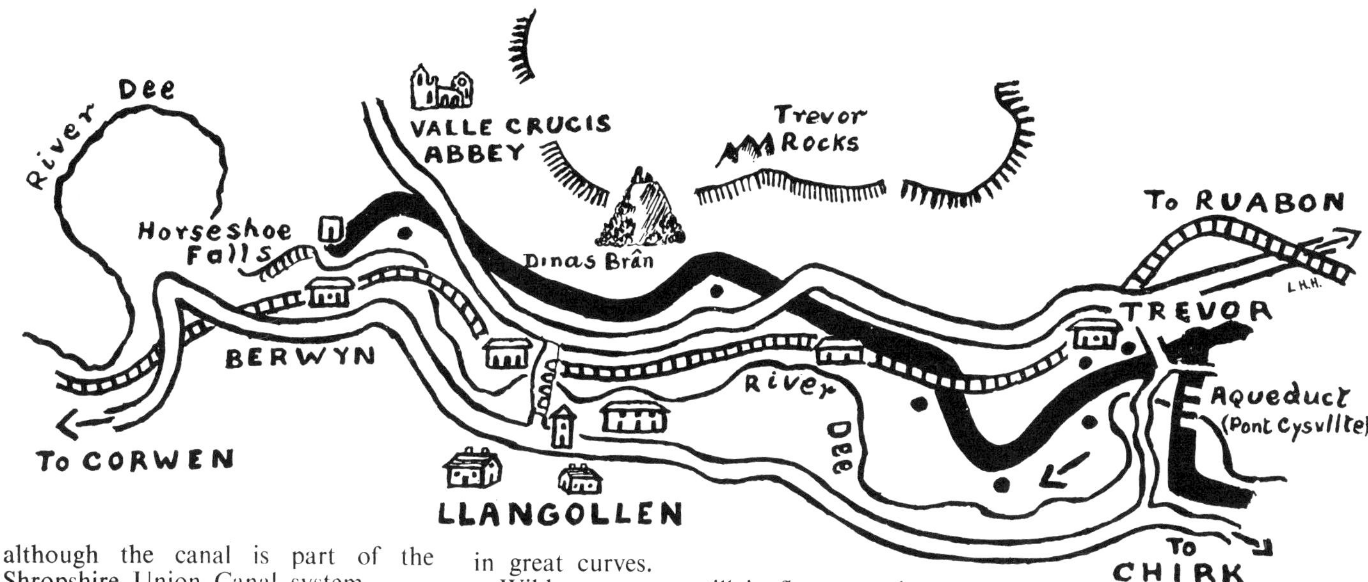

although the canal is part of the Shropshire Union Canal system.

This was the most beautiful walk of the series, and it would be hard to find more charming scenery than the valley of the Dee, between wooded hills and craggy mountains. The canal is high above the river, almost on the hillside, so I had lovely views all day.

I rejoined the tow path where I had left it on the previous day. It is at first on the right of the canal which, however, it soon crosses by a wooden bridge, after which the tow path is on the left—that is when facing towards Llangollen.

After passing an enormous boulder on the opposite side of the water, and beneath another wooden bridge, I had a spectacular sight of the great railway viaduct crossing the valley, and a lovely view of the Dee, below woods hanging on the hillsides.

The sound of the river, dashing over rocks, was very companionable. During the day's walk the Dee was generally in sight—near at hand, or winding off in great curves.

Wild roses were still in flower, and were hanging over the dry wall on my left. I passed the time of day with the man who tends this length of the canal. His nice black dog was guarding his coat and his lunch basket. I passed under several stone bridges.

Just short of the one numbered thirty-eight I had a fine view of the curious formation of the pink and yellow Eglwyseg rocks, and of Castell Dinas Bran, a conical hill, crowned with the ruins of a medieval castle.

Now the canal passes under the railway, which runs downhill on the left. Here is Sun Bank Halt—not very useful, since only one train for Llangollen stops here daily, and not even one in the opposite direction !

A piece of the canal bank, filled in with great stones, shows the place where some years ago the canal burst through its bank and carried away a piece of the railway below.

Immediately afterwards a mail train came on the line. The engine fell into the gap made by the rush of water, and both the engine driver and stoker lost their lives.

Very soon, road, railway and river are near together on the left, but far below. The traffic can be heard rather than seen, for it is hidden by the trees and undergrowth which cover the steep bank. Across the canal stands a house with flowery terraces descending to the water.

I next came to a very picturesque stretch, for, on the right, great limestone rocks, where ferns grow, rise high above the canal, and there is a view of mountains ahead.

Owing to the woodland hiding the valley I quite suddenly found myself above Llangollen, which is reached from the canal by a flight of steps. But I had decided to carry on as far as Berwyn, and to visit Llangollen on my way back.

From the head of the steps, a service of pleasure barges runs frequently in the summer months between Llangollen and Berwyn. I decided to try the experience of being upon the water, instead of beside it. These pleasure boats hold about twenty people. A roof protects from sun and rain but the sides are open.

The boats are drawn by horses in the old way. Our horse was very large, and the boy who bestrode him very small. He could only climb up onto the horse's back by making use of the chains to which the tow rope was fastened.

The boats coming from Berwyn seemed to have the right of way, for when we met one our boy would dismount, unfasten the tow rope from the horse and throw it into the water, retrieving it very dexterously with a hook, when the other boat had passed.

Besides this public service, a number of punts and canoes were travelling to and fro, upon the canal. It was delightful to glide upon the still water along the lovely valley. We had a glimpse of the slender metal bridge above the Dee, and of the railway station higher up.

From the canal terminus, near the Berwyn Hotel, a few minutes walk brought me to the so-called Horseshoe Falls, really a great semi-circular weir, not unattractive when a good rush of water is coming over, after heavy rain.

Here there is a new "Flow Meter" which checks the amount of water taken into the canal from the Dee.

On this side of the Dee is Llantisilio Church, beautifully situated, and, about a mile further, the lovely ruin of Vale Crucis Abbey, in exquisite scenery.

The railway and bus services from Berwyn to Llangollen, and on to Ruabon, are both on the other side (the right bank) of the Dee. On the whole I think that the best views are from the railway, which crosses the river between Berwyn and Llangollen.

On this occasion I returned to Llangollen by the tow path. It was quite remarkable how different the canal and the scenery looked when I was beside the water, instead of upon it.

Back at Llangollen, I descended the steps into the town. Llangollen is, of course, a beauty spot, and is also a splendid centre for walks and excursions. The four-arched bridge, built by John Trevor, Bishop of St. Asaph, in 1346, is one of the "Wonders of Wales."

Below the bridge the Dee foams over the rocks which form the bed of the river.

After a welcome tea at the Bishop Trevor tea rooms, at the foot of the steps, and opposite the bus stop and railway station, I returned by train to Ruabon, where I changed for Shrewsbury.

The distance from Trevor to Llangollen is about five miles, and another two miles on to Berwyn.

Seen from the Towpath......9

From Cheswardine to Market Drayton

by

LILIAN HAYWARD

THIS short walk of a few miles only is beside the main Shropshire Union Canal, near Market Drayton. The Midland Red Bus Company runs an hourly service from Shrewsbury to Market Drayton. It is an interesting run, through several pretty villages.

There is Shawbury, where Shropshire's best thatcher lives, and Hodnet, one of the prettiest villages in the county. Most of the houses here are half timber and some are still thatched. On the left there is a good view of Hawkstone Hill and, on the right, of Tern Hill.

Nearer Market Drayton is Salisbury Hill where, it is said, that the Earl of Shrewsbury and his followers encamped on the night before the battle of Blore Heath, 1459, one of the battles of the Wars of the Roses.

Blore Heath is only a few miles away, and here the Lancastrian Army was defeated. A cross marks the spot where their leader, John Audley, was slain.

From Market Drayton the Potteries Bus Service runs to Cheswardine on Wednesday, leaving Market Drayton at 9.30, 10.25, 12.40, 1.30, and 2.15; on Saturday at 12.55, 1.30 and 2.45 (these times should be confirmed, as usual).

Just outside the town, where the road crosses the river Tern, we pass into Staffordshire. In fact, we shall be in and out of Shropshire and Staffordshire during this walk. From the top of a hill, soon after crossing the Tern, there is an attractive view back to Market Drayton, the church standing up well.

The "Four Alls" Inn is passed. This has a quaint sign, depicting four characters: the Statesman, "I govern all;" the Bishop, "I pray for all;"

the Soldier, "I fight for all;" the Citizen, "I pay for all."

The bus should be left at Cheswardine Bridge, which crosses the canal, and where the walk begins. But as this is a short walk, you may like to walk a little way in the opposite drection, to see one of the few genuine old canal inns, still licensed — the Wharf Inn, near Goldstone Common. Here, refreshment can be obtained, and in the evening barges can generally be seen, drawn up to the wharf beyond the bridge. It is half-a-mile back to Cheswardine Bridge.

For more than a mile, in the direction of Market Drayton, the canal lies in a very deep cutting, between rocks covered with hartstongue and other ferns. In Spring there are masses of primroses. But it was in October that I took this walk. Traveller's Joy was foaming over the rocks, the trees were already changing colour, and yellow leaves were forming a skim upon the still water.

The two immensely high bridges under which I passed, give an idea of the great depth of the cutting, and the labour that must have been required to excavate the bed of the canal. It was the work of Telford, who seemed never to be daunted by difficulties.

This is the main line (if I may use the expression) of the Shropshire Union Canal, and here are to be seen barges, horse drawn, or propelled by engine, carrying cargoes to Norbury, Wolverhampton and further. Occasionally we may have the luck to meet with one of the gaily painted boats with a family aboard.

Beyond the deep cutting I came to

One of the Tyrley Locks.

Tyrley Lock and I was now in Staffordshire again. There are five of these locks, several of them close together. A carved stone on a quite picturesque red house, across the water, is inscribed "Tyrley Wharf," while another bears a date, A.D. 1857.

At Tyrley I was pleased and quite excited to see Mr. Rolt's "Narrow Boat," which I had last seen near Pont Cylliste in the Summer. "Ahoy Cressy," said I, not feeling sure whether this was the correct way to hail a canal boat a few yards away. "Hulloa," said Mr. Rolt, "you seem to have done a lot of walking by the canals since I saw you before."

I watched the Cressy passing through the locks—quite a business. Mrs. Rolt went on ahead to open them, and they are so close together that she had to scurry from one to another.

I noted the great length of the Cressy, as compared with her width, her many windows gay with flower boxes and cheerfully patterned curtains. The inevitable water-can, adorned with traditional designs, stood conspicuously upon the deck.

Cressy, with her crew of two, must have come right along the Ellesmere Canal to Nantwich, and so on to the Main Line Canal. She was now bound for Northamptonshire — a far cry indeed.

From Tyrley Bridge there is a field path, a short cut to the town, I, however, continued my way along the tow path. In contrast to having been at the bottom of a very deep cutting, after

I passed Tyrley, I found the canal to be high above the surrounding fields.

At one point, the canal passes over a road, far below. Forty steps lead down to it. Beside the road, the Tern is just visible through the trees — a pretty bit of scenery.

A few minutes walking brought me to a bridge over which the main road to Newcastle-under-Lyme passes, and here I left the canal. On the other side of the bridge were moored a number of strange-looking square white craft, which I was told are let to visitors in Summer.

At this bridge, I was back in Shropshire, Market Drayton being on the very edge of the county.

This was a very short walk of not more than four miles, so I filled up the afternoon by exploring Market Drayton, in company with a kind friend, who showed me round.

It is a pleasant old market town, on rising ground, with winding streets and old houses, some Tudor and many Georgian. Beyond the columned Butter-Cross, there is a market every Wednesday, when the stalls run the length of the long street.

The annual horse fair, "Dirty Fair," is the event of the year. The ancient Court Leet, which had been held for centuries, unfortunately died out some years ago.

The Church has a fine up-standing tower. It is said that Lord Clive, as a boy, got on to the top of this tower and then lowered himself on to one of the gargoyles, on which he sat astride, to the horror of lookers-on.

Almost in the churchyard is the house which was formerly the Grammar School, which Clive at-

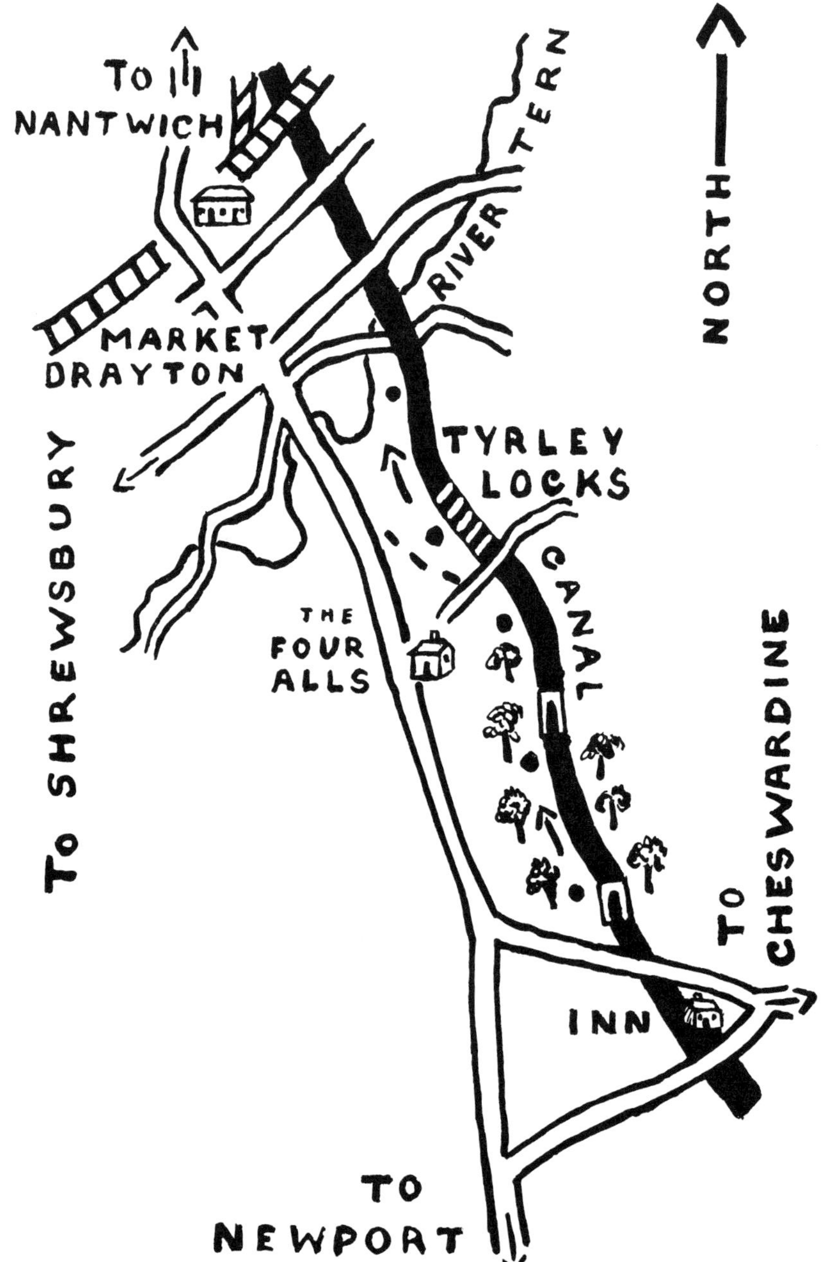

tended. He was the son of an impoverished Shropshire squire. His home was the Styche, in Moreton Say parish, a few miles away.

It is said that Clive used the first prize money which he received in India, to pay off his father's debts, and to redeem the old family home. He was buried in Moreton Say church.

Market Drayton church has several interesting features, including a beautiful West doorway. Inside I was especially charmed by a group of little heads round the head of a column. These include three wearing mitres and one a crown.

In the S.E. corner of the Sanctuary there is a good brass to Rowland Corbett (1580) who is depicted kneeling in a full-sleeved gown.

In a chapel on the North of the church is an incised alabaster slab, representing a Bulkely and his wife in sixteenth century costumes.

From the churchyard, on the edge of the crest, there are pleasant views into Shropshire and Staffordshire.

A deep cutting which shows the perseverance of Thomas Telford, the great civil engineer.

Seen from the Towpath......10

Our Waterways are in danger

by LILIAN HAYWARD

THIS is the last of the series. If any readers have walked along these tow paths, in fact or in fancy, I hope that I may have led them to realise the charm and interest of our canals.

We have not exhausted even those of the old Shropshire Union Canal Co. There is yet the link from Whitchurch to Hurleston Reservoir, and then down the main canal to Nantwich and Market Drayton.

Another much branched section leaves the Ellesmere canal, three miles south-west of that town, passing near Hordley and Bagley. Mr. Hyde (Ellesmere) runs a useful bus into this district on certain days. If any reader hopes to utilise this bus, I hope that they will have better luck in obtaining a time table than I had!

Beyond Hordley this branch passes near West Felton, and after a few miles comes again into charming scenery, near Pant, beneath a high bank, which in spring is white with hawthorn; and below Llanymynech Hill.

After this we are in Montgomeryshire, passing through hill country with delightful views all the way to Newtown. This branch is known as the Welsh Canal, although it is still part of the Shropshire Union system.

After the Burgedin Locks the canal lies in the beautiful wooded valley of the Severn, never far from the river, with grand views of the Breidden, and other heights.

The Severn Valley is quite as lovely

in its own way, as that of the Dee. Here and there are to be seen some of the beautiful old houses, once the homes of distinguished Welsh families. Penarth, near Newtown, is a gem of half timber work.

At Berriew, a show village, the canal crosses the sparkling little Rhiew by an aqueduct. Close by is St. Beuno's Stone, which traditionally marks the place where the seventh century saint first preached Christianity in the district.

Between Llanymynech and Newtown the canal is seldom far from road and railway, so it is not difficult to find a train or bus to take one to or from different points. There is a Youth Hostel between Montgomery and Newtown.

The Welsh Canal came upon evil days in 1944, when there was a burst. This could easily have been repaired, but it suited the Railway Co. which then owned this canal, to make this burst an excuse to abandon that part of it.

The destruction of this canal for practical use is being carried out in another ruthless way. In several places where roads cross the canal by the old hump-backed bridges, these have been lowered, and the water piped under the new road, thus effectually preventing the use of the canal for any form of water traffic.

This is the work of the County Council. Had they been more public-spirited they might have saved the canal and bridges by making long, gradually-sloping approaches. It is sad to witness the deterioration of what used to be one of the most beautiful water thoroughfares in the country.

Every few months we read of another section of a canal going out of use. The canals, like the railways, are now nationalised. One would have hoped that an enlightened government, which had the welfare of the people at heart, would have made more effort to maintain the canals, and bring back others into use.

They should encourage their use for business and pleasure. They might encourage sending goods by water, which is less costly than by train. They might promote the use of the canals for holiday cruising, or even organise excursions and cruises by water themselves.

In some sections there are agreements with the owners of land adjoining the canals that water should be maintained in them. But this is not universal. Moreover, it appears that to-day if any influential body finds its agreements or covenants inconvenient, it can get them nullified by a special Act of Parliament!

Those who are interested in the preservation of our canals and their tow paths should join the Inland Waterways Association (11 Gower St., London, W.1). This is the only organisation which concentrates upon inland water traffic, on rivers and canals.

The aim of this Association is "that every waterway in the country should be a well-kept navigation, freely open to all traffic of suitable dimensions, and at a standard rate of tolls."

Or, it would be helpful if sympathisers would join the Shropshire branch of the Council for the Preservation of Rural England (School Gardens, Shrewsbury), which has been making a spirited effort on behalf of the Ellesmere and Welsh Canals.

Union is strength. It is quite surprising what can be achieved by a number of firm-minded people banded together.

These articles have really been written from the point of view of the walker, rather than from that of the boatman. For easy walking, for loafing on an idle holiday, or for a family picnic, I cannot imagine any inland scenes more delightful than these tow paths.

It must not be forgotten, however, that should canals be definitely abandoned, the tow paths would also disappear, and no longer be available to ramblers.

Perhaps this is the moment to point out that there is no right of way on the tow paths. We are allowed to use them "on sufferance only,' to quote the notices.

Any hooliganism, or other damage done, might well result in the tow paths being closed to the public. As always, in such cases, the innocent would suffer with the guilty. The prohibition would no doubt include the anglers, who to-day so much enjoy their sport in the waters of the canals.

During these rambles I have been surprised not to see more artists at work with sketchbook or easel. For indeed there is an attractive picture at every bend; a mass of yellow iris or monkey plant, contrasting with the blue of sky and water; a church steeple; an old bridge of picturesque form, with a peep of hills beyond it; a waterside inn or farmhouse; or one of the great heavy-timbered draw bridges, which makes a picture by itself.

The story of the Ellesmere Canal....1

When £1,000,000 was subscribed in an afternoon for an enterprise now sadly declined

by E. A. WILSON, M.A.

ONE of the charms of England is that one may often find a fascinating history hidden behind a comparatively commonplace present in a great many towns and villages. The details of this history, however, may be far from easy to obtain.

In Ellesmere, the Canal is taken for granted. When I tried to collect information about its history, I gained little but boyhood reminiscences from "locals." However, at the Canal Depot I was loaned a Report headed:

Report
to
the General Assembly of
the Ellesmere Canal Proprietors,
held at the Royal Oak Inn,
Ellesmere, on the 27th Day of November, 1805,
to which is annexed
the Oration
delivered
at Pontcysylte Aqueduct
on its first opening, November 26th, 1805.
Printed by Order of the General Assembly,
the Right Honourable
The Earl of Bridgewater, Chairman.

This meeting was held during the year of the Battle of Trafalgar. The subject matter is set out under five headings. Under the first head—it is pointed out that "the Ellesmere Canal, rather than being one general Canal, is a system of Canals, distributed over that extensive and fertile country, which lies between the banks of the Severn on the South, and those of the River Mersey upon the North; and between the skirts of North Wales on the West, and the borders of Staffordshire on the East; a space of fifty miles in length and more than twenty in breadth.

"This Canal will unite the Rivers Severn, Dee and Mersey, and will open an inland navigation from the above mentioned districts to the Ports of Liverpool and Bristol.

"Although the commercial connexions, established by means of this Canal, will be far from inconsiderable, yet it is from fuel and from manure for agricultural purposes, that its most important and permanent revenue must be looked for."

This last paragraph needs explanation. When the first canals were built the Industrial Revolution was only just beginning, and the economy of England was essentially rural. Most of our population worked on the land, there being few large towns. In 1760 London had a population estimated at 700,000 and only Bristol, and possibly Manchester, had over 50,000 inhabitants.

Such industries as coal mining, iron smelting and pottery manufacture were trying to expand, but they were greatly hampered by transport difficulties. The general state of the roads was still terrible. They were often bad too for wheeled traffic so that pack horses were largely employed to take pottery and other goods to the nearest seaport, and to bring back raw materials. One can well imagine that breakages of pottery were numerous!

The price of coal rose steeply with the distance from the mine, and the Duke of Bridgewater employed Brindley to construct a ten-mile canal between Worsley and Manchester. Opened in 1761, it halved the price of coal in Manchester.

The "Bridgewater Canal" began a great boom in canal building. The industrialists saw a solution of their transport difficulties.

Nevertheless, the greater part of England was to remain rural for a long time to come, and many branches of the canal system, now derelict, finished up at some small country village.

Views of the Pontcysylte Aqueduct.

The canal in its cast iron trough and the tow path on the Aqueduct

This was the result of the country people needing lime (called "manure" in the Report) for their land and for building purposes, coal for their hearths and for "burning" limestone, slates for their houses and barns, and timber for building. The return traffic consisted of the produce of the farms, particularly cheese, for the towns.

In these days it is difficult to picture the enthusiasm over the proposed building of the Ellesmere Canal. To quote from the Report:

> "the paroxysm of commercial ardour of the memorable tenth of September (1792), can never be forgotten by the writer, who had the honour to be left to defend the hill near the town (Ellesmere) which gives its name to the undertaking, from the excessive intrusion of too ardent speculation: the books were opened about noon, and ere sun set near a million of money was confided to the care of the Committee. This wonderful act of public confidence was in a few days justified, by returning of all the money or value of shares not wanted for the works then projected, to those who had so liberally supplied it."

Broadly speaking the Ellesmere Canal system formed a cross with Ellesmere near its centre. The N.W. fork was planned to lead from the Mersey at Ellesmere Port (so called because it was the port at the seaward end of the Ellesmere Canal) across the Wirral Peninsula to Chester; then on to the coal mining districts around Wrexham and Ruabon; across the valleys of the Dee and Ceriog by aqueducts; and thus on to the junction at Welsh Frankton near Ellesmere.

It was intended that the canal would then proceed by the S.E. fork to join the River Severn at Shrewsbury. The S.W. fork was to be made to bring limestone from Llanymynech into the rest of the system and make connection with the Montgomeryshire Canal leading through Welshpool and Newtown into South Wales. The N.E. fork was to pass through Ellesmere and Whitchurch to the neighbourhood of Nantwich. Later the older Chester Canal was to connect it with Chester.

As so often happens the work did not go according to plan, although the proprietors were fortunate enough to acquire the engineering services of Thomas Telford.

He tackled the most difficult portion of the Canal almost at once; that concerned with the crossings of the rivers Dee and Ceriog. It was ten years before this work was brought to a successful conclusion, by which time the rest of the canal as it stands to-day was also finished. To quote:

> "this division of the Canal is composed of works more difficult of execution, than can perhaps anywhere be found within an equal distance upon Canal Navigation. From Chirk Bank on the south side of the River Ceriog to the north bank of the River Dee, is only four miles, but in this distance there are two very large Aqueducts, two Tunnels, nearly a mile of very deep cutting, and the remainder consists of benchings along very high steep banks."

The different "lines of Canal" were completed and adopted in the following order:

First. **The Wirral Line,** opening up communication from Liverpool and Manchester with Chester. They hoped that this would encourage the Chester Canal proprietors to revive the Canal from Chester to Nantwich where it would connect up with a proposed branch of the Ellesmere Canal.

This latter idea became an essential part of the whole system when surveys showed that it would be too difficult and costly to build the branch from Chester through Wrexham and Ruabon.

Second. **From Chirk to Llanymynech and Weston.** The aim of this line was to open up a communication between the colleries of the Ruabon neighbourhood and the Limestone Quarries at Llanymynech, and from these:

> "into Montgomeryshire in one direction and into the county of Salop at Hordley and Weston, in the other"

Originally this latter portion was intended to continue on to Shrewsbury, but the lack of funds prevented its completion.

Third. **From Frankton (upon the last mentioned line) by Ellesmere and Hampton Bank to Tilstock Park.** The main purpose was to extend the consumption of coal and lime in the surrounding country.

Fourth. **From Tilstock Park to Whitchurch, and the Chester Canal near Nantwich..** This would also supply coal and lime to the neighbourhood, connect up to the Wirral Line via the Chester Canal, and supply water to the Wirral Line which previously had to have water pumped into it from the tideway by means of a steam engine.

Fifth. **The Chirk, Pontcysylte and Llangollen Line.** There is always a difficulty in supplying the stretch of canal which runs at the highest level (known as the "summit level") with

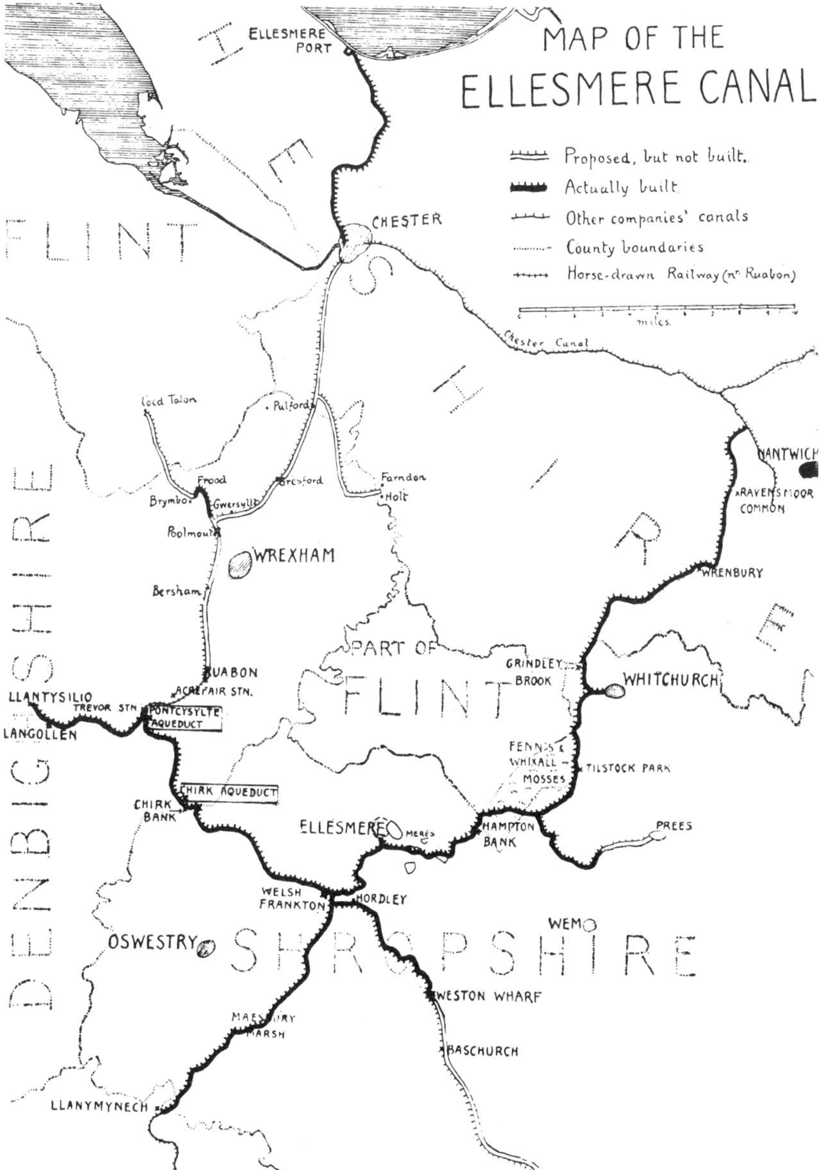

sufficient water. The "water line" led from the Pontcysylte Aqueduct as far as Llangollen where it connected up with the Dee at Llantysilio; the Dee itself came from Lake Bala where the Ellesmere Canal Proprietors had "works" controlling the amount of water let out from the lake into the Dee and hence into the whole of the Ellesmere Canal.

Beside this important reason, coal from the Ruabon Collieries was brought to the Canal at Pontcysylte by horse-drawn railway, limestone from Trevor, and slate from Irenant.

These reasons were considered important enough to justify the huge expenditure on aqueducts and tunnels which were among the first works started and the last to be completed.

To the scene was later added the competitor — the railway viaduct alongside the aqueduct

Check House and approach to first lock. Canal Tavern in background.

The third lock.

When peaceful Welsh Frankton was a busy centre with Wharves, Warehouses and a boat building Dock

by E. A. WILSON, M.A.

LEAVING Ellesmere by the Oswestry road, the road rises steeply for a considerable distance. It levels out again before reaching the highest point, and if conditions are favourable, you will have a grand view of distant hills on your right.

The climax is reached, when you reach the "Brow" by a small Congregational chapel. Here is a grand panorama, with the Breiddens descending steeply to the floor of a distant valley, flanked by wave after wave of impressive mountains.

Nearer at hand, the low lying land of the Welsh Marches has a softer appearance with green fields and hedgerows plentifully studded with stately trees.

In the sunshine of a summer evening it appears to be asleep, an impression which grows stronger as we descend from the Brow to the little village of Welsh Frankton (which in spite of its name, is in Shropshire).

Early in the nineteenth century this village was the meeting place of the two major portions of the Ellesmere Canal system, and there must have been many a merry party at the Canal Tavern between the crews of barges coming in with coal and slate from the Ruabon district, who had crossed the dizzy heights of the Pontcysylte and Chirk aqueducts, and passed through the equally eerie long tunnels, and the crews coming from the limestone quarries at Llanymynech.

Yet other bargees who met here may have come from Merseyside via the Wirral branch of the Ellesmere Canal; continuing by the Chester Canal to join the Ellesmere Canal again at Hurleston near Nantwich; passing over Whixall Moss and within a stone's throw of Colemere and Blakemere; on through Ellesmere, and so at last to Frankton.

Further boats came from the opposite direction from Weston Wharf loaded with timber from the Boreaton Woods and cheeses from the surrounding countryside.

Nor was all the traffic "one way" and as Frankton was the meeting place of all these journeyings it must have been a scene of constant activity in those days of canal pros-

perity.

During visits to this neighbourhood I have talked with a number of people, including the "lengthman" responsible for canal maintenance, and have tried to find out what purpose was served by the buildings and constructions which still remain, and any which have now disappeared.

The canal here has the form of a letter H lying on its side with the connecting link running north to south. Thus there is a northern junction and a southern junction. There are two bridges fairly close together, one on either side of the northern junction, to allow horses coming along the towpath from Ellesmere to cross over as the towpath changed sides here.

There used to be finger-post at the northern junction which had "to Newtown" pointing south; "to Chirk and Llangollen" pointing west; and "to Chester" pointing east. This has now gone.

At the wharf there was a warehouse, but this has been pulled down and the materials used to build a bungalow elsewhere. A crane beside the warehouse has also gone.

Along the link section there are four deep locks, the first two forming one unit with the lower gate of the first lock forming the upper gate of the second. Between them is a footbridge.

At the point where the locks begin, there is a small building known as a "check house" where a check was kept on the nature and amount of materials passing through the locks. To the left of the check house a lane descends steeply to the "stables" and the "lock-keeper's house."

Although the locks are no longer used, their stonework remains in sound condition to remind us of the craftsmen of bygone days. The wooden gates, being less durable, show signs of decay. The ironwork is well rusted.

Along this part of the canal there is a drop of about forty feet divided among four locks. Going south from the first pair of locks we have the stables and lock house on our left at a low level, and the old Canal Tavern high up on our right.

Mrs. Hyde lives here. The Hydes owned barges and did much canal carrying until comparitively recent times. With the passing of the canal traffic they diverted their energies to starting a bus service and to-day Hyde's Bus Services are well known in and around Ellesmere. Their headquarters are still at Welsh Frankton.

Perhaps the habits of the bargees of bygone days shocked the inhabitants into sobriety; there is not a single public house in the village to-day.

Then comes the third lock followed by a boat-building dock on our left. This can easily be missed as it has not been used for many a year and its entrance into the canal has been walled up. Only the newer quality of this stonework betrays its presence.

In this dock barges were repaired, and even built. Mr. Rolt's well known narrow boat, the "Cressy," was overhauled and refitted in this dock for a previous owner.

I could not get a satisfactory photograph owing to the vigorous weeds in and around it, hiding it almost completely from view.

There were two stone baulks to support a barge when the water was emptied from the dock by a sluice, and stone steps leading down into the dock.

The Canal Tavern at Welsh Frankton with steps direct from the canal to it.

The building just beyond was the home and workshop of the boatbuilder. Part of the workshop has been converted into extra rooms for the house, and practically all traces of the original equipment have been removed. However, I did come across a pair of bellows, seven feet in length, which had been used to blow the forge fire.

The fourth lock is followed by a stretch of canal almost completely blocked by rushes, reeds, sedges and other marsh plants before we reach the southern junction with the Weston-Llanymynech line.

If the neglect and desolation met with so far is somewhat depressing, what now meets the eye is far more so. On either side of the terminal weed-fringed pool, the canal has been drained completely. One can now walk for miles along the almost dry canal bed, in the direction of Weston.

Before nationalization of the waterways most of the canals had been bought by railway companies, chiefly to prevent competition with rail traffic. When there was a burst in this line in 1934, the owners decided not to repair it. The line of canal involved runs from Weston, through Welsh Frankton and on to Llanymynech where it joins the old Montgomeryshire Canal. This leads to Welshpool and Newtown, a total distance of 41 miles.

The canal bed, was sold to those through whose land it passed. Last year, not far from Weston, I met a bulldozer filling in and obliterating all traces of the canal, thus adding an extra strip of land to the farm. Several farmers have done this.

Nearby, a canal bridge was being demolished. Unlike modern railways and roads, the old canals and their bridges have mellowed with age into almost natural features of the countryside, and one cannot but regret their passing.

The half destroyed bridge showed up the method of construction, namely a tunnel-like arch, *only one brick thickness*, covered over with rubble and having a brick parapet on either side. The fact that these bridges stand up to modern lorry traffic is a tribute to the sound design and workmanship.

Old workshop and forge bellows at former boat building site at Welsh Frankton.

The 1805 Report makes it clear that it was intended to continue the line from Frankton to Weston to join the River Severn at Shrewsbury, and it states:

> "These Lines proceeded on with considerable rapidity, and were partially opened in 1797; but this exertion exhausted the funds allowed by the Proprietors, ... when these lines were opened, and a wharf, four lime-kilns, a public house, stables, a clerk's house, and weighing machine, constructed at Weston, it was judged adviseable to postpone proceeding further in this direction, because the adjacent country could, from this central point, be supplied with fuel and manure (i.e. lime), to a distance which came into competition with the coal and lime from Ketley, Steeraway and Donnington Wood.
>
> "The execution of these works has been of considerable advantage to the country and the Proprietors, by having promoted the sale of great quantities of lime and slate at Weston wharf; but notwithstanding the navigation has for some time past been extended to within two miles of the Ruabon Collieries, there has scarcely any fire coal ever yet been sold at that wharf."

On the way to Weston to see what remained of the buildings mentioned in the Report, I visited Hordley Wharf. Three years ago the basin contained some water; two years ago it was marshy; now it was dry, the sun-baked clay bed being cracked into the semblence of crazy paving.

I walked dryshod over this floor where the shells of the freshwater mussel still remained and dwarf clubrush and water-pepper grew, and climbed up on to the wharf whose surface was beautified by an abundance of the tiny white-flowered Whitlow Grass.

Looking westwards, one's gaze met a dreary stretch of dried grass-covered canal; eastwards there were trees to cover up its derelict condition. Did barges once unload here? It was difficult to believe that this completely deserted spot was ever of any importance.

Further on, the road crossed and recrossed the canal several times, until at last I reached Weston Wharf. Here the canal passed under a bridge into the basin. Once more I found the basin dry and grass covered.

Hard by the bridge there was a warehouse built right up to the edge of the basin so that boats could load or unload directly from it. A short distance away were the four limekilns built as one block, one of the openings now serving to shelter a small car.

I was conducted up some steps to the top of the kilns where I had an overall view of the buildings. Yes! there was the Canal Tavern, and next to it the Clerk's House, now serving as ordinary dwelling houses, and both looking extremely sound although over 150 years old.

The Stables were still there and so was the Weighing Machine. A crane which once stood on the wharf seemed to be the only object which had disappeared.

In 1800 this "central point" was a busy terminus. To-day all is changed and even the canal itself is fast disappearing.

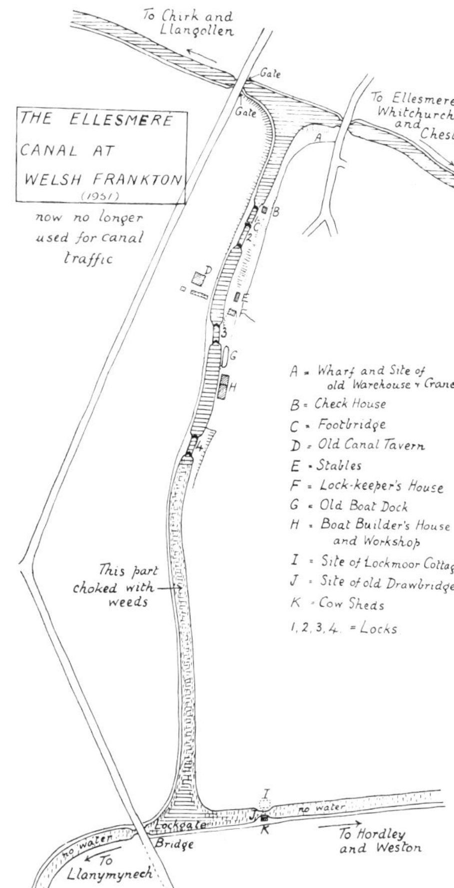

Where stout British oak lock gates are still being made

by E. A. WILSON, M.A.

"Beech House," Ellesmere.

Canal Depot at Ellesmere. Note supports for the overhead crane. The closed door leads to the dry dock. Note also, maintenance barge on left. Beech House in background.

THE Report of the Ellesmere Canal Proprietors for 1805 states: "the Committee have judged it necessary to erect a Canal Office at Ellesmere, being the most central point in the Navigation. This will contain a Committee Room, an Office for Accounts, and another for Plans, etc., also apartments connected therewith for the resident Accountant Agent, and the resident Engineer."

A glance at the map of the Canal shows clearly that Ellesmere was the natural focal point of the system. The "Office" was built as a house of ample size and dignity, sited at the junction of three branches of the canal.

It commanded a view in all three directions from three windows of a Committee Room which formed the ground floor of a projecting wing of circular shape.

"Beech House," the present name, has been a private residence for longer than any local people can remember. To-day, it is converted into flats.

Adjacent to Beech House is the Canal Depot, as old as the canal it serves. It is pleasant to be able to depart from our tale of neglect and decay and say that. These works are much alive and employ some twenty men in servicing an even larger area of canals than before nationalization.

Here I saw lock gates weighing several tons being made from English oak and strengthened with iron.

There is a dry dock for repair of barges and other vessels. It contained a weed-cutting boat during one of my visits.

Dominating the scene is the massive structure for the overhead travelling crane. The uprights consist of tall square-sectioned baulks of timber supporting equally large beams fixed horizontally. One such structure runs down each side of the yard until it overhangs the water of the canal, producing the effect of a colonnaded avenue.

Until recently, power was obtained from an old Cornish engine of great age. It is now preserved in a museum. The engine used to-day is rather unusual, being adapted from a locomotive, the cylinder and piston being mounted vertically. It came from Crewe.

Many a barge was built or repaired at the Canal Depot in bygone days, but there is now little demand for such work, and only occasionally does one see a boat of any kind tied up.

More likely than not, it will be one of those ugly hulks built of riveted iron plates and used for maintenance and repair of canal banks. One hardly knows whether to laugh or cry on finding the name "Venus" on the bow of such a blatantly utilitarian construction.

I had high hopes of being able to examine old documents, accounts, maps and other material at Beech House which would throw light on the history of the Ellesmere Canal; but they have long since disappeared.

At the Canal Depot I was loaned a copy of the 1805 Report, by Mr. Hughes, the Canal Inspector for the North Western Division of the Docks and Inland Waterways Executive, and I am most grateful for his help. Inquiries about further documents to Divisional Headquarters brought the reply that they would be grateful if I discovered them! I wonder if any reader has such documents?

Leaving the Canal Depot and following the short branch which finishes at Ellesmere Wharf one passes under a footbridge, known locally as the "White Bridge." This is really a "roving bridge" built to allow horses to cross over at a junction to the branch line towpath.

Just beyond this bridge there *was* a two storey building containing a boat-dock. I have not yet had a satisfactory explanation of

Terminal Basin and wharf at Ellesmere. Note old type crane and warehouse on left with old lettering on end wall.

The Boat House and dock recently demolished at Ellesmere. Note "roving bridge" and shelter for "stop-planks."

the original use of this building, but one story was that it was used as a small boat-building site.

In later years it was the property of the Bridgewater Estate and housed the estate barge. A boat club in Ellesmere used it as their boathouse and Mr. Wainwright told me that members used to row to Frankton and Weston, or past the Meres to Whixall Moss. Later Mr. Caseley took over the running of the club.

Recently the club ceased operations, and on Whitsunday, 1951, I saw workmen stripping the tiles off the roof. Rushing home for my camera I photographed this building just in time! To-day the dock is almost completely filled with bricks from the demolished building.

Passing on, the canal soon widens out to form the terminal basin at Ellesmere Wharf. An ancient crane stands at its head, and although no longer in use, is still in good preservation with all its gears intact. Standing by it, and looking back towards Beech House, we can see a large warehouse on our left, with the following information on its end wall:

SHROPSHIRE UNION
RAILWAYS AND CANAL COMPANY
GENERAL CARRIERS
TO
CHESTER, LIVERPOOL
NORTH AND SOUTH STAFFORDSHIRE
AND
NORTH WALES

In 1846 the Shropshire Union Railways and Canal Company was formed by the amalgamation of the Ellesmere and Chester Canal, Birmingham and Liverpool Junction Canal, Shrewsbury Canal, and Montgomeryshire Canal. The warehouse, however, is probably much older than this.

Beyond the warehouse is an open area formerly a coal wharf. Then there is a smaller warehouse with a smaller crane in front of it.

On our right Clay's Ironworks or "The Bridgewater Foundry" used to stand. This not only made ironwork for the canal but served a larger area and even exported its manufactures. To-day the site is occupied by a modern, clean-looking depot of United Dairies. The fleet of lorries is housed in an extensive modern garage built to the left of the old warehouse, but some lorries stand by the Canal Wharf as well. I asked if they

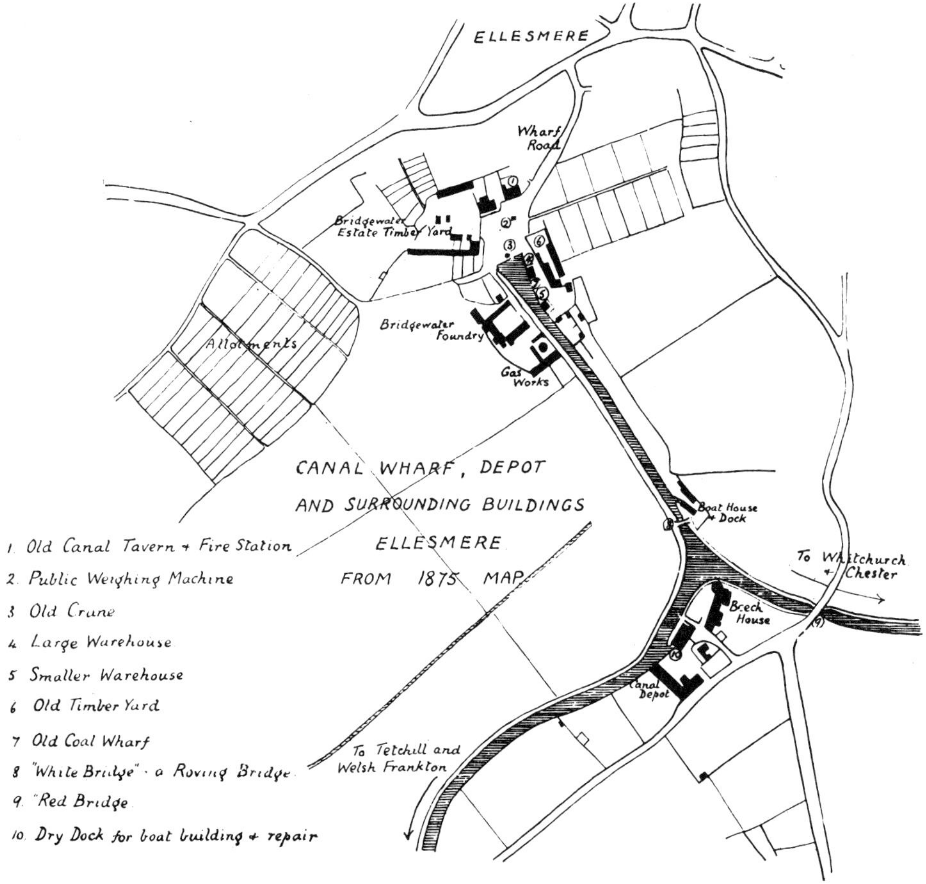

could be temporarily removed to enable me to get a photograph. My request was readily granted, with the result shown.

An activity of the United Dairies factory is cheese-making. Cheeses used to be made at farms all over Shropshire and Cheshire, and up to the first world war there was an annual Cheese Fair at Ellesmere when there would be vehicles stretching from the Church to the Market Place, their owners waiting for their cheeses to be assessed and bought.

Many cheeses were dispatched by canal to Liverpool and other towns en route, and this commodity often received priority treatment and was despatched by "fly boats."

These were barges of lighter build which travelled fast, having the right of way at bridges and locks, sometimes sounding a horn to give notice of their approach. They worked continuously, having a double crew, two men to work by day and two by night, only stopping at locks or when changing horses.

Near the Wharf stands the Public Weighing Machine, no longer used. Also near, stood that indispensable adjunct, the Canal Tavern. In part this now serves as a dwelling house, but an adjoining part became the local fire station in the days of the old manual fire engines, and the wide doors common to fire stations can be seen in the photograph.

There were stables at the Tavern, but there was far from enough stabling accommodation around the Wharf and many horses had to be taken to other taverns in the town.

Many people in Ellesmere remember when the Wharf basin was crowded with barges awaiting their turn to unload, but I have been unable to find a single photograph of that colourful scene. Thus we must each form our own picture of this busy centre of the Ellesmere Canal system in the past, helped by such tantalizing glimpses from the Report as:

"He (Thomas Denson, resident engineer) is to manage all the operations in the Company's Yard at Ellesmere, and give proper directions to the tradesmen employed by the Company, to measure their work, determine their prices, and examine and sign their accounts, and deliver the same to the general accountant."

The story of the Ellesmere Canal,,,,4

How £459,461 was spent – They complained of labour shortage and high costs in 1805 !

by A. E. WILSON, M.A.

MUCH interesting material can be obtained from old account books—the goods made and sold, the places from which materials were obtained or to which they were dispatched, the firms and people involved.

Thus, with great interest I studied the

Two tunnel views

"FIFTH HEAD" of the 1805 Report, which gave a "General Statement of the Ellesmere Canal Receipts and Expenditure," from the start to December, 1805.

One must remember that figures given in these articles represent the value of money in 1805. Making allowance for this, it can be seen that the undertaking was colossal, as the total expenditure up to 1805 is given at £459,461.

Before the Canal could be built, it required an Act of Parliament. Every extension proposed needed a further Act. Surveys had to be made before the actual route was chosen. Landowners along the proposed route might haggle over the price of their land. Some might refuse to sell.

The proposed canal might divide their land and they might insist that the Canal Company should build an "accomodation bridge." to connect the two parts. Many interests who feared loss of revenue, such as turnpike road trustees, had to be placated.

All this preliminary work needed the offices of skilled lawyers, and the Report gives the following items:

Messrs. Potts & Leeke. Clerks, Solicitors and Law Agents to the Company, for Salary from Jan. 1, 1794, to Oct. 1, 1805	£3,060 17 6
Ditto, professional charges for attending Parliament, and business done for the Company, from April 12, 1791 to Oct. 29, 1805 ..	£3,156 6 0
Ditto, for sundry travelling expenses, expenses of witnesses and law payments, from April 12, 1791 to Oct. 29, 1805.	£2,836 14 3

Business connected with Parliament is mentioned in the item:

Messrs. Whites of the House of Commons, for fees, expenses, etc.	£1,802 1 6

There are a number of entries connected with the surveying, making of plans and drawings, and similar matters:

John Duncombe, for salary in making surveys, sections, plans, attending Parliament, and acting as resident engineer to the Company from 1791 to

June 30, 1803, including all travelling and other expenses allowed him. ..	£4,402 2 5
Davies and Jebb, land surveyors, for making original plans and books of reference, also the subsequent ones, attending Parliament at sundry times, valuing land and damage during the execution of the work, and making plans of the Canal as completed, including expenses	£3,010 9 10

The total under the heading of "Land and Damages" is given as £40,744 5 9, one item being the payment of £625 to Hughes and Dod, millers.

Thomas Telford, the designer and architect of the whole scheme is termed the "General Agent," and he is mentioned as follows:

Thomas Telford, as general agent from Oct. 11, 1793 to Dec. 25, 1805, as per his accounts, including his expenses upon the lines of the Canal	£3,639 12 5

Thomas Denson, who, on the completion of the Canal became the Resident Engineer is also mentioned:

Thomas Denson, paid by the Treasurers from Jan. 1798 to Christmas, 1805 ..	£1,782 3 2

Canal shares were usually in denominations of £100. A deposit was payable when the shares were first taken. As the canal was built, "calls" of so many pounds per share were made on all shareholders until their shares were fully subscribed.

If parts of the canal completed brought in a good revenue, the shareholders were usually eager to pay up when "calls" were made to finance new construction.

The Napoleonic Wars, however, caused a labour shortage and a rise in the cost of materials. "Calls" had to be made frequently. Some shareholders could not or would not pay up, or delayed payment as long as possible. How common this was may be realised from the following entries:

Collecting calls and sundry small payments by Treasurers	£2,800 8 0
Expenses in collecting calls and in attending Parliament, paid by Thomas Telford, as per his accounts	£1,193 7 10½

We find that although 3,330 shares were issued, only a small proportion of the total value was ever collected on 247½ shares.

Most works were let by contract or paid for by "measure and valuation." Very little was done by direct labour. For instance, for the Wirral Line we find:

Weston and Fletcher, for works performed partly by contract and partly by measurement and valuation	£32,679 5 5
Works performed by measurement and valuation, under the direction of the Company's superintendent, John Telford	£28,552 8 2

One interesting item is: John Wilkinson for engine, etc. £1,576 0 6. The Wirral Line was opened in 1795 before the other branches, and had to maintain its water supply by raising water from the tideway by a steam engine operating a pump. This ceased to be necessary when the whole system was completed.

Total expenditure on the Wirral Line up to 1805 is given as £68,405 19 0.

The accounts show that the Chirk, Llanymynech and Weston Line, not including the Chirk Aqueduct, cost £55,691. We are given the cost of certain materials used—bricks, stone, timber, quicksetts, lime, posts and rails. There must have been much woodwork as Thomas Thomas, carpenter, was paid £1,351 8 4.

Samuel Betton seems to have contracted to do most of the Frankton to Tilstock, including the Prees Line. It included the difficult crossing of the Mosses and his bill came to £39,995 11 9, out of a later cost of £49,194 odd.

John Lee was paid £234 6 6 for "wooden bridges, etc." John Simpson, £5,848 17 1 for the tunnel near Ellesmere and the usual canal bridges.

The famous Shrewsbury ironmaster, William Hazledine, comes into the picture for the first time with a modest bill for £70 18 8 for "castings for swivel bridges, etc."

The Tilstock Park to Nantwich portion cost £40,186 7 0.

The most spectacular part of the whole canal, that beginning at the Chirk Aqueduct and finishing at Llantisilio, shows only two entries under "Chirk Aqueduct":

Simpson and Hazledine	£19,055 0 10
William Hazledine, iron work	£1,843 11 11
Total	£20,898 12 9

Telford "cradled" canals over aqueducts in an iron trough. In this particular case, the section running over the aqueduct had an iron bed, but stone sides. The Chirk aqueduct is much wider than the Pontcysylte aqueduct, in consequence.

Beside the two aqueducts, there were two long tunnels, and nearly a mile of deep cutting, and the rest ran alongside steep banks.

John Simpson, an expert in tunnel construction, received £3,258 4 3½ (why the ½d.?) for making "tunnels, bridges and culverts"; Thomas Davies £6,119 10 2 for "cutting, etc."; while together Simpson and Davies received £20,554 13 1 for "tunnels and deep cuttings by measurement and valuation." The total of this difficult portion was £45,527 9 0½.

At the southern end of the Pontcysylte Aqueduct there is an earth embankment, 1,500 feet in length, and seventy-five feet

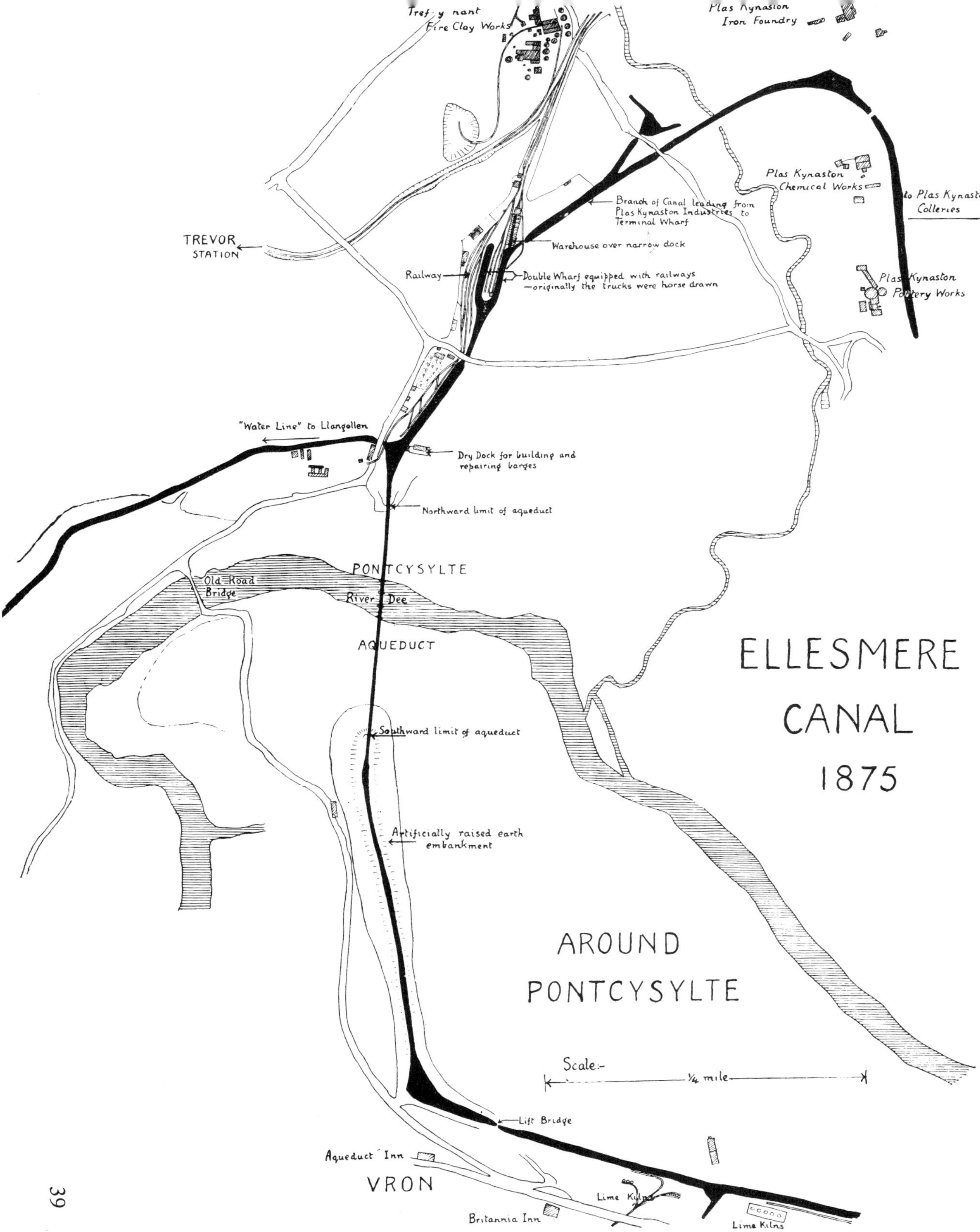
ELLESMERE
CANAL
1875
AROUND
PONTCYSYLTE
Tref-y-nant
Fire Clay Works
Plas Kynaston
Iron Foundry
Plas Kynaston
Chemical Works
to Plas Kynaston
Collieries
Plas Kynaston
Pottery Works
Branch of Canal leading from
Plas Kynaston Industries to
Terminal Wharf
Warehouse over narrow dock
TREVOR
STATION
Railway
Double Wharf equipped with railways
—originally the trucks were horse drawn
"Water Line" to Llangollen
Dry Dock for building and
repairing barges
Northward limit of aqueduct
PONTCYSYLTE
Old Road
Bridge
River Dee
AQUEDUCT
Southward limit of aqueduct
Artificially raised earth
embankment
Scale:-
¼ mile
Lift Bridge
Aqueduct Inn
VRON
Lime Kilns
Britannia Inn
Lime Kilns

high at the northern end. Raising this embankment was no mean feat and Mr. Davies was paid £8,570 15 8.

This magnificent structure, opened to navigation in 1805, is a memorial to the genius of Thomas Telford, its designer, and the men whom he employed.

William Hazledine was paid £17,284 17 5½ for the ironwork. The accounts of John Simpson, mason, who was responsible for the erection of the arches, along with James Varley, were:

James Varley, for masonry	£2,001 19 10½
Simpson and Varley, ditto	£10,445 14 7
John Simpson, ditto	£8,520 1 1

William Smith was paid £46 15 10 for a model of one arch, and the grand total was £38,498 10 11.

The heading, "railway" seems out of place in the Napoleonic Wars. It does refer to a rail *way*, but the trucks were horse drawn. Roads were then so bad that much greater loads could be hauled by horses when the vehicles ran on rails, and it was usual in colleries and ironworks to transport goods for short distances by this method. The report says:

> "From the north end of the Pontcysylte Aqueduct the Canal is continued for about three hundred yards, and there terminates in an extensive basin, formed so as to afford a double wharfage with iron railways.
>
> "From this basin an iron railway has been laid and carried near Plas Kynaston Stone Quarries, through the Acrefoir Colleries, and is to be continued to the Ruabon Brook. The length of the railway will be about three miles."
>
> The cost is given as £5,888 14 0.

The last portion of this line led the Canal along the north side and nearly parallel with the River Dee for some six miles. Going through Llangollen it finished by making connection with the River Dee at Llantisilio.

Large passenger barges still make the journey from Llangollen to Llantisilio, and a better setting for a pleasure trip would be hard to find.

The cost of this "water line," i.e. the line which supplied the Ellesmere Canal with its water supply, was £10,724 12 10.

An isolated canal, only three miles long was built near Wrexham, called the Frood Branch. It began "at some extensive Coal and Ironworks in that neighbourhood, and is no great distance from some Lime Kilns."

The cost was £8,221 15 2.

It has already been told how Canal Taverns, Houses of Clerks, Stables, etc., were erected at various centres, and a considerable number of lime-kilns and wharves along the Canal. The cost of these items was £9,968 13 4½.

What a tremendous spate of activity is represented by these accounts! The whole of the scheme took only ten years to complete, in an age when manual labour was the rule and there were few labour saving devices.

The Story of the Ellesmere Canal—5

About "Narrow Boats," "Bridge Holes," "Fly Boats" and the strenuous art of "Legging"

by E. A. WILSON, M.A.

CANALS in England were built in a rather haphazard fashion by many independent authorities. There was no uniformity. The cost was governed by width and depth, so they were built as small as their particular purpose would permit.

The Ellesmere Canal was mainly "agricultural" and did not form part of any main trunk route. Thus, considerable stretches are both narrow and shallow.

Many bridges were required to carry roads and connect farm lands divided by the canal. To save cost, the canal was narrowed so that the "bridge-hole" allowed one boat at a time.

Locks were expensive, and along most of the Ellesmere Canal were only wide enough for one boat.

This narrowness controlled the design of the barges. The only way to make larger barges was to extend their length. Even this was limited by the length of the locks, so that a more or less standard type of barge, known technically as a "narrow boat" came into use on the smaller canals.

At first, a towpath was necessary, as narrow boats were either towed by men ("bow hauling") or, more usually, by animals. A single horse was often used. "Fly-boats," the express boats, used for quick delivery, had two, as did "packet boats" for passenger traffic. Sometimes two donkeys were substituted for a horse.

When a canal branched, the horses had to cross over the canal to get on the towpath of the branch, and a "roving bridge" was built over, or near to, the junction.

Tunnels were very costly, and to save expense they were sometimes made with no towpath. The horses were taken over, or around the high ground by special paths, while the narrow boat was propelled through the tunnel by the men lying on their backs and pushing against the walls with their feet. This was known as "legging."

The walker along the towpath finds a wooden shelter housing a number of stout wood planks at many bridge holes. Inspection will reveal an iron-bound groove in the masonry on each side of the canal.

The planks are known as "stop planks." They fit into the grooves to form a dam. A section of the canal can thus, be temporarily emptied for repairs. Nowadays, the shelter is often replaced by a long box made up of square concrete sections, cemented together and closed by iron doors.

When I first came to the Ellesmere district, I was puzzled by the old, derelict kilns at various points, though there is no limestone in the district. Lime was needed for agriculture in the rural areas of Shropshire. Limestone was quarried at Llanymynech and brought by narrow boat to these kilns where it was burnt to provide lime.

I have found lime-kilns as follows: two kilns at Colemere, where the canal runs at a higher level than the mere, so that the top of the kilns is level with the canal, enabling the limestone to be loaded directly into the kilns; a block of three kilns at Hampton Bank; a block of four at Weston Wharf; and at Quina Brook, where the Prees branch ends, a block of five, facing the Wem-Whitchurch road, with three more in the side of the same bank. Those facing the road are so hidden by trees and bushes, that most people must pass them by unnoticed.

When designing canals, allowance had to be made for exceptional conditions, such as heavy rain, which might lead to overflowing. Sluices were provided where the canal ran along an embankment, allowing excess water to be run off, but an additional safeguard was the "storm-overflow," a broad stone (or concrete) runway leading from the canal bank just above the normal level to a lower level.

There is such an overflow at Colemere where the storm water can flow from the canal into the mere, and a larger one, adjacent to an accommodation bridge at the beginning of Fenn's Moss.

From time to time, the banks need attention and the addition of puddled clay. Along the Prees branch there is an extensive clay pit, from which clay has been obtained for

The Colemere Sluice.

The Ice Boat

Hampton Bank Wharf - lime kilns are situated behind the bank.

Isolated warehouse not far from Quina Brook along neglected portion of the Prees branch.

"Roving Bridge" and house at junction of Prees branch with the main canal. Monkey-musk flowering along bank.

this purpose for many years. The trucks are filled with clay and hauled by wire rope up to the wharf, where the clay is loaded into maintenance barges. Sometime back they took their loads to repair the bank along by Fenn's Moss.

Some wharves are simple landing stages, as at Hampton Bank and Colemere, but others have warehouses and then there was usually a crane to aid the loading and unloading of the barges. Two such cranes survive, though not used, at Ellesmere Wharf; the crane at Weston Wharf has disappeared; and at Welsh Frankton both warehouse and crane have gone. However, at Bettisfield, I recently saw both a warehouse and crane in use.

The flow of water in canals is very slow and they sometimes get frozen over. This was serious when canals were the main arteries of transport, and special "ice-boats" were built.

There is one of the old wooden ones aground at Ellesmere. There is a slatted wooden deck with a strong post at each end, supporting a stout, horizontally-placed iron bar. The bow was protected by iron.

Several men stood on the deck on either side of the iron bar, and gripping this firmly, rocked the boat vigorously from side to side to break up the ice, while six horses strained their hardest to keep the boat going forward. The man at the tiller endeavoured to steer a direct course.

Mr. L. C. T. Rolt, in his book *Narrow Boat*, paints the scene vividly: "Two gangs of breakers made a spectacular arrival at Banbury. . . . . Long before they came in sight their approach was heralded by a grinding and crashing sound, then round the bend came the sweating horses keeping a fine pace, as though entering into the spirit of the adventure.

"Finally, the boat itself appeared, rolling almost gunwale under from the efforts of her heaving crew, who, ruddy-faced from the cold wind and strenuous labour, seemed oblivious alike to the jets of icy water which spurted from overside and the dirty puddle which slopped to and fro beneath their feet. At the tiller stood an elderly lengthman, balancing first on one leg, then on the other as he endeavoured to keep the bucketing craft upon some semblance of a course."

There used to be a certain amount of passenger traffic on some of the canals, and although I doubt whether this applied to most of the Ellesmere Canal, we have a most extraordinary "museum piece" in one of the Duke of Bridgewater's original packet boats, the "Duchess-Countess."

It has now come to rest beside the canal in the peaceful fields not far from Welsh Frankton, and serves as the home of an elderly, but very active, person who once travelled on her.

Coming upon the boat one evening, its owner, seeing my interest, courteously invited me aboard. I was shown pictures and cuttings, and was able to learn something of the boat's former history.

The Duke of Bridgewater, who built the canals which started the "canal age" ran packet boats by which passengers were conveyed at the rate of a penny a mile. He often travelled on them himself. The "Duchess-Countess" was probably built 150 years ago. For nearly half a century she plied daily between Stockton Quay and Manchester, carrying passengers, her daily trips starting at 6.0. a.m. and finishing promptly at 7.0. p.m.

She was manned by a crew of three, the captain, the mate and the "jockey." Although only of six foot beam, her length allowed her to carry thirty passengers and provide for their refreshment. Drawn by relays of four horses, she travelled at six miles an hour.

Boats travelling outwards from Manchester had the right of way over those approaching the city, but the "Duchess-Countess," the pride of the canal, had the right of way in either direction. Any boat disputing this right was summarily dealt with, for at her bows there was a large S-shaped knife, which cut through its opponent's tow rope in a most cavalier fashion!

Passengers became fewer as the "railway age" progressed, and at last this proud old aristocrat of the canals was reduced to the ignominy of carrying goods, cattle and poultry and bringing back fustian.

Her daily trip was finally abandoned in 1915. She was tied up at Stockton Quay for a time and then was given a fitting burial when she was taken to Runcorn and submerged with some other boats in the Big Pool.

However, this was not the end. After being submerged for 18 years she was refloated again in 1934, when a retired Warrington man refitted her and with two friends, set out to "explore" the Ellesmere and neighbouring canals.

One of these friends was the present owner who finally brought her to her present resting place and lived afloat in her for some time. Although the bones of the old aristocrat were still strong, time had taken its toll, and she began to leak. It was war time, labour and materials were scarce, so with the aid of friends the owner dragged the boat out of the water on to the bank.

Here she lingers on. For now much longer? One can hardly tell. But surely this must be one of the most interesting links in that great and glorious story of British enterprise which produced great men and great works. Let us hope that the "Duchess-Countess" will not be allowed to succumb a second time, but will find a fitting resting place in time to come, where all may draw inspiration from her story.

Shropshire Canal's Economy Squeeze

AN 1822 COMMITTEE

by Edward A. Wilson

AT the meeting of the Canal Proprietors held at Ellesmere on the 27th of August, 1821, it was resolved:

"That it is desirable to examine into the annual expenditure of supporting and maintaining the Ellesmere and Chester Canal, for the purpose of ascertaining whether the same can be reduced."

The resolution was carried, and a Committee of three appointed to inspect the whole line of the Canal. They were given free access to all documents and powers to ask any questions of employees. They fixed on the 25th of June, 1822, as the date on which they were to meet at Chester, but only two members of the Committee, Mr. Joseph Lee and Mr. Exuperius Pickering, junior, turned up, their report stating "that they were disappointed in their expectation of meeting Mr. Freme." However, these two members decided to carry on and were provided with a horse-drawn boat which took them first along the nine miles of the Wirral Line to Ellesmere Port, and then back again to Chester.

There were extensive works at Ellesmere Port—which, by the way, came into existence as the "Port of the Ellesmere Canal" —where the Tide Basin and its Lock connected directly with the wide estuary of the River Mersey and not, as now, with the Manchester Ship Canal. A similar Tide Basin and Lock were present at Chester, where the Wirral Line of the Canal locked down into the estuary of the River Dee.

A thorough examination of all these works, and of the bridges along the Wirral

The Beeston "Iron Box" Lock.

Line, was carried out by the Committee and any repairs needed were noted down. Two of the bridges, one on the road from Chester to Mollington and the other at Caughall, had been rebuilt in cast-iron; they must have been badly in need of repair as there had been several fatal accidents to passengers in the "Packet Boats" while passing underneath. The mention of Packet Boats is interesting, as it reminds us of the fact that passenger services were run on this and other canals at this period.

Next came the inspection of the twenty miles from Chester to Hurleston Junction near Nantwich where many improvements had recently been carried out. The lock at Beeston Brook had been a constant source of trouble as it had been built upon a quicksand, and water carrying the sand away with it, had undermined the foundations; this had necessitated an expenditure of £200 during the previous year. It remained a perennial source of expense until 1828, when Thomas Telford had the lock replaced by one which was constructed of cast-iron plates, flanged and bolted together, virtually forming a water-tight iron box. This proved to be a satisfactory solution to the problem and Beeston Iron Lock is still in use today.

From Hurleston Junction the Canal passed southwards through Wrenbury, Whitchurch, over the desolate Whixall Moss, to Ellesmere and on to Welsh Frankton. The Committee reported that many of the former wooden lockgates had been replaced by gates made of cast-iron. The first of these to be introduced was at Welsh Frankton. The Report states that the four upper gates and one pair of lower gates there were made of cast-iron, as were all the gates at the three Aston Locks, and those of the upper lock at New Marton. In their concluding remarks we read:

> "The Company is indebted to Mr. Telford for the introduction of Iron Gates, the general adoption of which your Committee strongly recommend, being confident they will ultimately prove the best and cheapest."

Surviving lock gates at Welshpool with characteristic ground-paddle mechanism as generally used on the canal.

These cast-iron gates were supplied by William Hazledine, the Shrewsbury ironmaster, from his Plâs Kynaston Foundry, which had been set up by him to supply the iron work for the Pontcysyllte Aqueduct at the beginning of the century. Today not one iron lock-gate remains on this Canal, although I saw the last remaining pair on the neighbouring Montgomeryshire Canal at Welshpool last summer, where they form the upper gates of the Welshpool Lock, adjacent to Severn Street Bridge which crosses the canal nearby.

Continuing their inspection the Committee passed over Chirk Aqueduct, through the quarter mile long Chirk Tunnel, and over the astonishing Pontcysyllte Aqueduct where the Canal is contained in an eleven feet ten inches wide cast-iron trough with arches, supported on eighteen slender pillars, crossing the River Dee at a height of 126 feet. Then came the "Water Line" to Llangollen, and finally Llantisilio, where some of the water from the River Dee was side-tracked into the Canal by means of a horse-shoe shaped weir, now called the Horseshoe Falls. They reported that this weir had been partly carried away by the previous winter's floods and was being rebuilt.

An interesting section of the Report deals with the prices of labour and materials, and a most incongruous note is struck by such a statement as the following:—

> "Upon the works north of Chirk Bank the wages have been *reduced* from 12 shillings to 10 shillings per week(!).

Overseers of districts had their wages reduced from 15 shillings to 13 shillings per week.

While pointing out that the lowest wage paid by the contractors on the nearby Holyhead Road was 10 shillings per week, the Report concluded this section with the words:—

> "Your committee are aware that if, by a moderate addition to their wages, a judicious selection of workmen can be obtained, it is, in many instances, economy to offer it. But as a general scale of

wages they are of the opinion, with reference to the present state of the markets, that 9 shillings per week ought not to be exceeded."

By carrying out these reductions in wages it was hoped to save £53 per annum. £40 of this they proposed to use to increase the £80 allowance already made to the Clerk to the General Agent — a clear case of "unto everyone which hath shall be given: and from him that hath not, even that he hath shall be taken away from him"!

Comparing the expenditure per mile of the Ellesmere and Chester Canal with that of other Canals, the Committee found:

"in no one instance is the expenditure of any other Canal so little, and in some it is more than double"

Thus, in a spirit of optimism and mutual congratulations, the Report finishes.

Warehouses at Ellesmere Port built on arches over lower canal basin.

Ellesmere Canal Is To Begin A New Lease Of Life

by E. A. WILSON

THE Ellesmere Canal, which was originally built to carry limestone and coal to kilns for the production of lime, will begin a new lease of life in September.

It will be used by the Mid and South-East Cheshire Water Board to carry water from the River Dee at Vroncysylte (near Llangollen), to Hurleston Reservoir.

Until recently it had always been the practice of large towns needing pure water for its inhabitants, to take their supply from a mountainous area far removed from any likelihood of contamination by industrial effluents or even by cattle.

These sources were either natural lakes or artificial ones formed by damming up valleys. They were connected to the towns by long and expensive pipelines. Lake Vrynwy in the Welsh mountains supplies water to Liverpool in this manner.

It is now possible to use river water for the same purpose because of the development of purification plant which allows us to produce "potable" water, i.e., water fit to drink, from "raw" river water with comparative ease.

The Mid and South-East Cheshire Water Board needed increased supplies of water and hit upon the ingenious idea of using the Old Ellesmere Canal to solve their problem.

In this way no special pipeline would have to be constructed as the canal would serve as an open pipeline. It is estimated that something like three-quarters of a million pounds has been saved.

An agreement to carry out this scheme was reached with the British Transport Commission, the owners of the canal, in an Act of Parliament passed in 1954.

By this agreement the B.T.C. will be responsible for the conveyance of the water from Vron to Hurleston Reservoir and the cost of the necessary canal maintenance.

The Board will be responsible for a pumping station at Vron which will pump up an extra 8,000,000 gallons of water a day into the canal from the River Dee. It already carried over 6,000,000 gallons a day from the river to Llantisilio. They will also be responsible for the purification and storage works at Hurleston, and pipelines to distribute th- "potable" water to Crewe, Sandbach, Middle wich and other towns in the area.

The works at Vron are now completed. The pumping station is only a few hundred yards from Telford's great Pontcyslyte Aqueduct, but fortunately somewhat masked by the trees which clothe the steep banks at this spot.

The much more extensive works at Hurleston Reservoir which include a reconstructed inlet, raw water pumphouse standing in the reservoir itself, sedimentation tanks, rapid gravity filters, chemical treatment building, administrative block, sludge settling tanks and lagoons, pumphouses and pipelines, are expected to to completed by the late summer of this year.

So the old Ellesmere Canal has been reprieved and will remain open indefinitely.

The reprieve came only just in time as now no commercial craft navigate even the once prosperous Wirral Branch. The great warehouses and basins at Ellesmere Port are empty and silent.

They gave birth to a vigorous and rapidly developing town which has passed through youth, maturity, old age and is now quite dead.

According to Thomas Telford, who was responsible for the construction of the Ellesmere Canal, its main purpose was to promote the interests of agriculture by providing the farmer with lime for the improvement of this land.

This was done by conveying limestone from the quarries, and coal from the collieries to various vantage points along the canal where lime-kilns were erected for the production of lime.

Slate, building stone, road stone, stand, timber, grain, malt and general groceries were also carried.

There is no doubt that the canal proved a great boon to the surrounding countryside. The canal company did not carry goods themselves. but made a charge on the tonnage carried by private traders of 1½d. or 2d. per ton per mile.

Some idea of the traffic carried may be obtained by an inspection of the tonnage returns for 1809, which amounted to £3,411 13s. 9d. for limestone and lime carried, and £2,016 14s. 4d. for coal.

Pumping station on the Dee in sight of Telford's 150-year-old Pontcysylte Viaduct.

The first portion of the canal to be constructed was the nine-mile length running from Chester across the Wirral Peninsula to Ellesmere Port.

It was made of broad dimensions and later connected up with the broad Chester Canal. This length was extremely successful, and, beginning to operate in 1795, was soon earning useful dividends.

Not only was there much carriage of commercial goods, but a flourishing passenger boat service was in operation; and Ellesmere Port soon built up a reputation as a bathing resort !

Very large warehouses, basins, wharves, a boat-building slipway and other installations were erected to cope with the ever increasing traffic and even as late as 1843, after the "Railway Age" had begun, a new and larger basin was built at a cost of £100,000.

Many railways bought up competing canals and then neglected them deliberately, but the Ellesmere Canal was more fortunate.

Although acquired, with neighbouring canals, by the Shropshire Union Railways and Canal Company in 1846, the new company's attempt to drain the canal and run a railway along its bed failed, and they then leased the canal to the London and North Western Railway Company.

About 1860 the great railway companies were acquiring territory in the fierce competitive struggles then going on.

The G.W.R. had a line going from Shrewsbury to Chester, and further east the L.N.W.R.'s line was progressing from Crewe through Whitchurch to Shrewsbury.

Independent companies in Wales were beginning to join up their lines—later to be amalgamated into the Cambrian Railway System.

Before the Welsh railways were built the Ellesmere and Montgomeryshire Canals formed the only commercial route southwards from Oswestry to Welshpool and Newtown.

As neither the G.W.R. or the L.N.W.R. had a railway supplying the area, the L.N.W.R. maintained the Ellesmere Canal in an efficient state in order that it might compete with the Welsh Railways.

The policy ensured the survival of the Ellesmere Canal for almost another century, although with ever diminishing returns.

Its useful life continued for a time after the First World War, but with the development of modern road traffic fewer and fewer boats used it, and the railway company (now the L.M.S.) spent less and less on maintenance.

When there was a breach in the bank in 1935 near Welsh Frankton, they gave this as their reason for draining this part of the canal, and closing the rest of the Llanymynech Branch, the Weston Branch, and the whole of the Montgomeryshire Canal to navigation.

They were able to get rid of the Weston Branch competely, and much of it has been sold and bulldozed in and bridges demolished.

On the Llanymynech Branch several bridges have been lowered and filled in, the water being piped through the resulting embankment. Near Pant nearly three-quarters of a mile of the canal has been piped.

The Wirral Branch still continued to carry a certain amount of commercial traffic, but the remainder of the canal from above Llangollen, through Ellesmere and Whitchurch to the Chester Canal at Hurleston Junction carried none. Apart from the few maintenance boats and the rare pleasure there were no boats on it at all.

Why then was this portion of the canal not closed to navigation like most of the remainder? The reason was that it served as a "feeder" to convey water from the River Dee at the Horseshoe Weir, Llantisilio (near Llangollen) to the Chester Canal and the Wirral Branch, and also to the Middlewich Branch of the Chester Canal which joins the Trent and Mersey Canal.

Where the Ellesmere Canal joined the Chester Canal at Hurleston Junction a large reservoir had been constructed in 1829 to store water against periods of drought.

To convery water for this purpose and to control its rate of flow adequately a certain amount of maintenance was essential.

The passage of even a few boats is beneficial in keeping a clear channel. But maintenance boats had still to be used and to operate boats locks had to be maintained in a workable condition. Periodic weed cutting, dredging, and strengthening of banks was also necessary.

Some industrial concerns, such as the United Dairies at Ellesmere needed large quantities of water (most of which was returned to the canal) for condensing purposes; and farmers had come to rely on the canal for watering their cattle.

However, the British Waterways was not a philanthropic institution, but a nationalised concern using public money. The revenue from the Ellesmere Canal was now almost nil, but the maintenance costs were about £30,000 per year.

If it was to be kept open new sources of revenue had to be found, and one proposal was to encourage pleasure boat traffic by more positive measures.

The British Waterways chose certain canals for this purpose and published attractive booklets full of information useful to holiday makers. The first to be published was entitled "Cruising on the Llangollen Canal" This measure has certainly been successful in attracting many more pleasure boats. Canals are also popular with fishermen, and they contribute a small amount of revenue as a result of their activities.

All these measures, even if successful, were quite inadequate to offset the heavy maintenance costs.

So the beauties of Wales around Chirk and Llangollen, with Telford's magnificent aqueducts will be preserved for the enjoyment of many future canal voyagers.

A more mundane consideration is that much of the water on which they travel will finally disappear down the throats of the inhabitants of Cheshire, or to be used for washing, bathing and other purposes. Part of the water rates they pay will help to maintain this 164 year-old canal for many years to come.

How the old Ellesmere Canal played its part in the Lime Business

By E. A. WILSON, M. A.

THE old "Ellesmere Canal" (later to become part of the Shropshire Union Canal system) was built primarily with the needs of farmers in mind. As Thomas Telford wrote in 1800, "this is probably the first instance, in Great Britain, in which the extension and improvements of agriculture have been the principal motives that

A Lunar landscapee? No, just old limestone workings at Llanymynech.

Old-time kilns at Colemere.

led to the forming of a navigable canal of so great an extent and where manufactures, commerce and the trade of large towns have been only secondary considerations."

The almost universal fertilizer in those days appeared to be lime, and a branch of the canal ran to Llanymynech, where there were extensive limestone quarries; there were further quarries at Vroncysylte, near the famous Pontcysylte Aqueduct; and yet others at the Trevor Rocks close to Llangollen.

To aid the distribution of lime, a great many limekilns were built at strategic points along the course of the canal, and limestone from the several quarries was sent to be "burnt" in them—along with coal from the Ruabon and Chirk area.

The first limekilns I came across along this canal were in the woods at Colemere, not far from Ellesmere. Here the canal runs at a greater height than the mere, and the tops of the two kilns are thus level with the canal, allowing easy loading and unloading. The two kilns form one unit and are well preserved.

Since then, I have found further blocks of kilns which can be arranged in a regular sequence. At Hampton Bank three kilns; at Weston Lullingfields, where the Weston branch terminated, four; at the termination of the Edstaston branch, known as Quina Brook, five (with three more in a side bank); at Vroncysylte a huge block of six very large kilns.

It is interesting to note that the Weston branch was planned as part of the main line between Ellesmere Port and Shrewsbury, but when Weston was reached in 1797 "this exertion exhausted the funds allowed by the proprietors, and left very considerable arrears . . . it was judged advisable to postpone proceeding further in this direction, because the adjacent country could, from this central point, be supplied with fuel and manure (i.e. lime) to a distance, which came into competition with the coal and lime from Ketley, Steeraway, and Donnington Wood."

Likewise the Edstaston branch was planned to reach Prees, but again financial considerations caused it to halt in the heart of the country, and the eight lime kilns at Quina Brook prove the great need for lime in that area.

I have some bills headed "Ellesmere, Maestermyn and Quina Brook Lime Company." The block of two kilns by Maestermyn Bridge has literally been razed to the

The old Haulage Drum at Pant, which lowered trucks loaded with limestone down to the canal by an iron chain.

Horse traction still in use on Pen-y-graig quarry tramway, taking limestone to the tarmac works by the canal at Vroncysylte.

ground, and not a stone remains. However, old Ordnance Survey maps and Railway Plans show the site of this and other former kilns.

A visit to the places where the limestone was quarried shows us what important commodities limestone and lime were in the early days of the canal's existence. At Llanymynech and its near neighbour, Pant, there are extensive quarries high up in the limestone escarpment.

From them three self-acting inclined planes descend, the descending trucks loaded with limestone pulling up empty trucks.

Although these quarries are now forsaken and deserted, their impressive size is evidence of their former importance. Practically all the mechanism concerned with haulage has long since vanished but, surprisingly enough, there is a haulage drum at Pant which has survived. It is a huge wooden barrel with a number of iron wheels inside to give it the necessary strength. The stout wooden spindle of this barrel is supported between two massive limestone walls, and a considerable length of the original iron chain is present around the outside of the drum, as well as the broken brake mechanism. The whole structure is almost hidden by the growth of bushes and shrubs during the intervening years.

A walk down the incline from this point brings us out just above the now dried-out portion of the canal, where there are five more lime kilns.

The story had a different conclusion at Vroncsylte. Here limestone was brought to the canal from the lofty Pen-y-graig Quarries by a succession of three self-acting inclines, with horse traction on the level stretch between the first and second inclines and again at the bottom. Some of the limestone was taken down directly to the canal, the tramway going underneath the road; other limestone went to the top of the kilns at road level for conversion into lime.

The lime kilns are now no longer used, but the works survive because it was found that the quarried rock was suitable for road making, and the product now produced is tarmac, made by crushing the rock to a suitable size and coating with tar.

There is yet another use of limestone, and that is its use as a flux in iron smelting. Limestone from the Trevor Rocks at Llangollen descended a steep incline down to the canal at Sun Trevor Bridge, and from here travelled all the way to the iron furnaces on the Humber Arm of the Newport Branch of the Shropshire Union Canal.

Limestone or lime no longer travels along the old Ellesmere Canal, but in its day it made no mean contribution to the fertility and prosperity of the land through which it flowed.

The Ellesmere Canal is to enter upon a New Career of Usefulness

by JOHN DAVIES

THE Llangollen branch of the Shropshire Union Canal runs for some 46 miles from Hurleston, 5 miles west of Crewe, to Llangollen. It enters Shropshire near Whitchurch, passes south of Ellesmere and finally leaves Shropshire near Chirk. It is now planned to use this canal to carry water from the Dee to Hurleston for the Mid and South East Cheshire Water Board. This use as a conduit will enable it to be kept open for navigation.

It was this waterway which caused a complete revolution in my attitude to canals and occasioned a whole series of delightful experiences, not the least of which were connected with "messing about in boats." Before coming to live a few miles from it, at a point roughy half way along its length, canals had always been associated in my mind with gas works and heavy industry, the sort of thing that one sees when travelling by rail between Wolverhampton and Birmingham.

It was therefore with surprise as well as pleasure that a tantalisingly short stretch of water between reedy banks was noticed from the Whitchurch-Ellesmere road. Inevitably, one fine spring day the car had to stop while the tow path, the bridge and what was "round the bend" was investigated. On another occasion we made a trip there with the deliberate intention of stealing a yellow flowering water plant for a garden pool, and were actually in the act, holding the ankles of a friend, who, with his shirt sleeves rolled up, was pulling at the roots, when two cyclists appeared on the tow path. We were in imagination already in the dock waiting for the magistrate's sentence, but as they approached they turned out to be not policemen but plump young ladies who merely advised us to let go of his legs.

These walks became a regular feature until, inevitably, the idea of a boat arose.

It is not difficult to believe, when one wants to believe, the advertisements assuring one that with "X's" kit of parts, only a kitchen knife and a corkscrew are needed to complete a handsome 11 ft. dinghy, 18 ft. cruiser, or what have you. Perhaps a veil had better be drawn over the days following the arrival of the dinghy kit. Suffice it to say that building it proved quite beyond my powers and well within the capacity of two 15-year-old schoolboys who not only built it for me but even managed to be polite about my attempts to help them. They did it in ten days and it took me a fortnight to paint it. To those wanting advice on the building of a boat, I would merely say that the paint on the hull covers mistakes better than the varnish on the decks, and that gap filling glue will in fact fill surprisingly large gaps. One can even feel smug about this as the marine glue is said to be stronger than wood. Meanwhile, an obliging local engineering firm made a trailer which the head of the firm and I designed between us, my contribution to the discussion being, 'Yes, good idea," repeated whenever he paused for breadth.

My brother and his wife came out for the launching, and his is the dramatic photograph taken as the three of us eased "Water Rat" into the water. His intention was to take successive photos as it sank lower and lower into the water, ending with one of the concentric ripples as it went to the bottom; but then he thought I had built it. However, it did not sink, and I bravely stepped in.

It did not sink.

Our wives got in. Amazingly, it did not sink.

By now my brother was reassured and he put one foot gingerly into the boat. But now no-one was holding it and the bank quickly receded and the angle of his legs from acute became right-angled and then obtuse. Afterwards, when he was safely in the boat, his wife was heard to remark, almost in awe, that she had not previously realised the extent of his knowledge of Anglo-Saxon.

The rowing caused some difficulty, even after we realised it was better not to sit facing the bow, for the canal in places is quite narrow. Fortunately the inevitable collision between the zig and the zag was with a wall of reeds and this facilitated our study of marginal plants. The study of underwater plants only began in earnest when we bought an outboard motor and in some sections had to raise it to unwind yards of greenery from the propeller every few yards.

Once when it stopped without this encumbrance the petrol tank was found to be half full of a green solid, afterwards discovered to be the result of putting premium petrol in a brass tank, and not a new kind of algae.

It was on this expedition that we had a fracas with a swan. While the motor was working we had passed two swans, the female sitting contentedly on the nest and the male swan standing passively by. On the way back with the motor passive the male swan became most aggressive and chased the boat with wings beating, causing us nearly to overturn in our attempts to evade them. Even this had its compensations, however, for the feathers left in the boat were added to others we had picked off the water to be sent to an artist friend who makes quill pens.

We found the use of the motor to have some drawbacks. First it annoyed the fishers along the bank, causing the herons to fly away and the humans to give us the blackest of looks. Secondly it caused a quite terrifying vibration in the boat. This may have been due to the fact that we had inadvertently bought a motor with four times the power recommended. Or it may have been a result of our bland assumption that the wood we could find no place for in the boat, and had turned into a garden seat and other furniture, was sent in error.

Perhaps it is better to row than to motor for then one can hear the "plops" of the little pike, and the "tappy lappy" of the ripples against the bow. Then again one is all the more ready for the picnic on the tow path, often accompanied by a chat with one of the locals, perhaps a peat cutter from the Whixall Moss, who tells you of the "ethers", found to be adders, there, or more rarely with another traveller by canoe or cabin cruiser.

How nice it is to learn that the future of this canal is assured, that its quiet pleasures will yet remain for those who value them.

This old Canal Packet is ending its days as the "Home of a cheerful philosopher"

by H. CLAYTON JONES

The Duchess Countess, last of the Canal Packets.

HOUSEBOATS, from the luxuriously appointed craft on the Thames, the Norfolk Broads, and in our sheltered coastal creeks, to the humbler concerns in more modest backwaters, provide homes for a large number of people, but it must be unique to discover a cheerful old philosopher living alone in an aged craft that was once an aristocrat of the canal. To-day, reminding one of Noah's Ark resting upon Mount Ararat, the last remaining packet passage, or passenger boat stands high and dry upon the banks of the Shropshire Union Canal at Lower Frankton and provides a home for its owner who has lived alone aboard her since 1936. This boat, the Duchess Countess, was built over a century and a half ago and was one of the Duke of Bridgewater's original packet boats, in fact the name of the craft recalls the titles in the Bridgewater family.

The Duchess Countess in her hey-day conveyed passengers at the rate of a penny a mile between Stockton Quay and Manchester, the return journey occupying thirteen hours. Having a beam of six feet she was long enough to accommodate thirty passengers together with their goods and a day's refreshment. In the days when roads were often very indifferent canal travel was quite popular, for the canal packet boats were considered to be more comfortable and cleaner than the stage coaches, and their progress was certainly smoother though they rarely exceeded the speed of six miles an hour. The Duchess Countess had a crew of three, the captain, the mate, and the "jockey," and she was drawn by relays of four horses. Boats travelling from Manchester claimed the right of way over those approaching the city, but the aristocratic Duchess Countess had the right of way in either direction. To demonstrate her superior status she carried a large "S" shaped knife at her bows. If any boatman contested her right of way his towrope was promptly severed. It is interesting to note that it is recorded that there were occasions when passengers were "seasick" aboard the Duchess Countess.

With the rapid development of the railway the prestige of the canal packet boat declined, and when the conveyance of passengers became uneconomic the once proud Duchess Countess was demoted to the menial job of carrying cattle and poultry into Manchester, returning with cargoes of fustian. In 1915 the old craft did her last regular voyage and she was taken to Runcorn where she was ingloriously scuttled with other boats in the Big Pool there. This, however, was not the final scene in the life of the old boat that had served the public so well, for after being submerged for eighteen years she was refloated and used by her new owners for exploring the Shropshire Union and neighbouring canals. One of these explorers was the present owner of the Duchess Countess. For some years the skipper of the veteran craft kept her moored close to the small dock at Lower Frankton and from there he made voyages in different directions, a favourite journey being over the Welsh Border to cross the famous aqueducts at Chirk and Vron Cysyllte. The Duchess Countess was

Comfortable living quarters.

the last boat to navigate the Montgomery Canal before it was closed. In the war years the old boat began to leak badly, and as repair work was out of the question, a team of willing villagers dragged her ashore, where she still remains beside her native element where she once proudly sailed.

Although beacned amid delightful pastoral scenery the site of the old boat is a somewhat lonely one, but it must not be supposed that the skipper envies those living in more conventional surroundings, for he is perfectly happy with his wooden walls and his own satisfying philosophy of life. A small range serves him for heating and cooking, the mellow light of an old oil lamp lends a cosy atmosphere to his cabin, a wireless proclaims the doings of a less tranquil world, and a small workshop and a collection of well chosen books provide pleasant distraction.

The old fellow is a discerning reader who numbers Stevenson among his favourite authors, and when, in his shirt sleeves, he displays the skull and cross bones tattooed upon his forearm, one is tempted to look into the darker recesses of the boat and to see the one-legged old sea cook of *Treasure Island* or listen for the harsh voice of his parrot. Neither does this modern Peggotty find life uneventful, for his mild adventures include fishing unfortunate wayfarers out of the canal, and an encounter with an infuriated bull that got a horn wedged in the iron-work of the boat. In war days the local Home Guard, to which the old skipper belonged, made use of his boat when engaged in tactical exercises, after being warned of course that their language must not conform to standards usually associated with inland waterways.

As with the rapid development of rail travel the life of the canal packets was comparatively short there is nothing except a model in Lancaster Museum and a few contemporary engravings—except the Duchess Countess of course—to give us an accurate idea what these boats were like. It is inevitable that sooner or later, if left to the mercy of the elements, the timbers of this gallant old craft will disintegrate, so there is a movement on foot to ensure her permanent preservation, but until that is done she will remain the home of her present owner as a novel and interesting feature on the peaceful Shropshire landscape flanked by the quiet canal.

(Since this article was written preliminary steps have been taken for the preservation of the old packet boat.)

Maesbury Hall Mill from the old canal inlet

Maesbury Hall Mill once had its own Fleet of Narrow Boats on the Shropshire Union Canal

by E. A. WILSON, M.A.

THE little Morda brook has provided the power needed for working a number of water-mills along its course during times past. To-day most of these have ceased to operate and in some cases the buildings have gone, the millponds have been drained, and, in one case at least, a former millpond now serves as an allotment.

There is a notable exception to the extinction of such rural industries in this district. It is the Maesbury Hall Mill of Messrs. A. and A. Peate Ltd. Situated near Maesbury, it has a lovely setting hard by the now derelict portion of the Shropshire Union Canal. The mill itself is a strictly utilitarian structure, the modern additions which dominate the buildings having corrugated iron exteriors; but for all that, it is imposing in its size and proportions, and is a model of efficiency.

The best viewpoint is from the canal just by the bridge. A wide, funnel-shaped offshoot of the canal here narrows down to a point from which it used to continue as a small channel just wide enough to accommodate a single narrow boat on its way to Mill Wharf; the boat had to return stern first. To-day the narrow channel is filled in, but the view from the canal over the wider portion, framed on either side with woods, with meadows occupying the intervening space up to the Mill, and more trees forming the background, presents a most satisfying picture.

Built on the site of a former malt kiln, Maesbury Hall (Corn) Mill has been a family concern of the Peates since 1846. It was worked by Mr. Edward Peate until 1862, when the present owner's grandfather, Mr. Andrew Peate, purchased it and operated it until his death in 1872. His widow then let the mill for the next fifteen years until her eldest son, Andrew, was old enough to take over in 1887. Her second son, Albert, joined Andrew in 1892, and their partnership continued until 1927 when Andrew retired.

The present managing director, Mr. Frank Peate, joined his father when his uncle died in 1931; and on the death of his father in 1941 took over the management of the Mill himself.

Originally the Mill was a watermill fed from the Morda brook, and this source of power is still used, although it is now supplemented by electrical power. A few hundred yards north of the Mill, at the hamlet of Newbridge, there is a weir and sluices which enable the miller to direct the water from the brook into a substantial millpond; and from thence by an artificial channel to the Mill itself. After it has done its work by driving a water turbine the water is led back to the Morda brook by a further cut which is culverted underneath the canal. Thus, a short length of the brook is temporarily cut off while the miller makes use of this source of water.

No doubt a water wheel was first installed, but now a vertical water turbine is worked by a 24 feet fall of water. Progressive modernisation has characterised the Mill throughout its long history, and the stone grinding of wheat gave way to steam driven roller mills in 1890, and these were remodelled in 1908.

A fire in 1911 did considerable damage, gutting the provender mill, warehouse and screen room, but this only proved a temporary check as they were all rebuilt before the year was out. The following year saw the addition of a new silo of 700 tons capacity; and a new warehouse was built eight years later.

Electrification of the Mill took place in 1933 and soon afterwards the storage capacity was increased by extending the old silo and constructing a new one. Another improvement was the provision of a dryer house to deal with home grown wheat. This all arrives at about the same time, and in too great a quantity to be dealt with immediately. After a typical English summer it may be

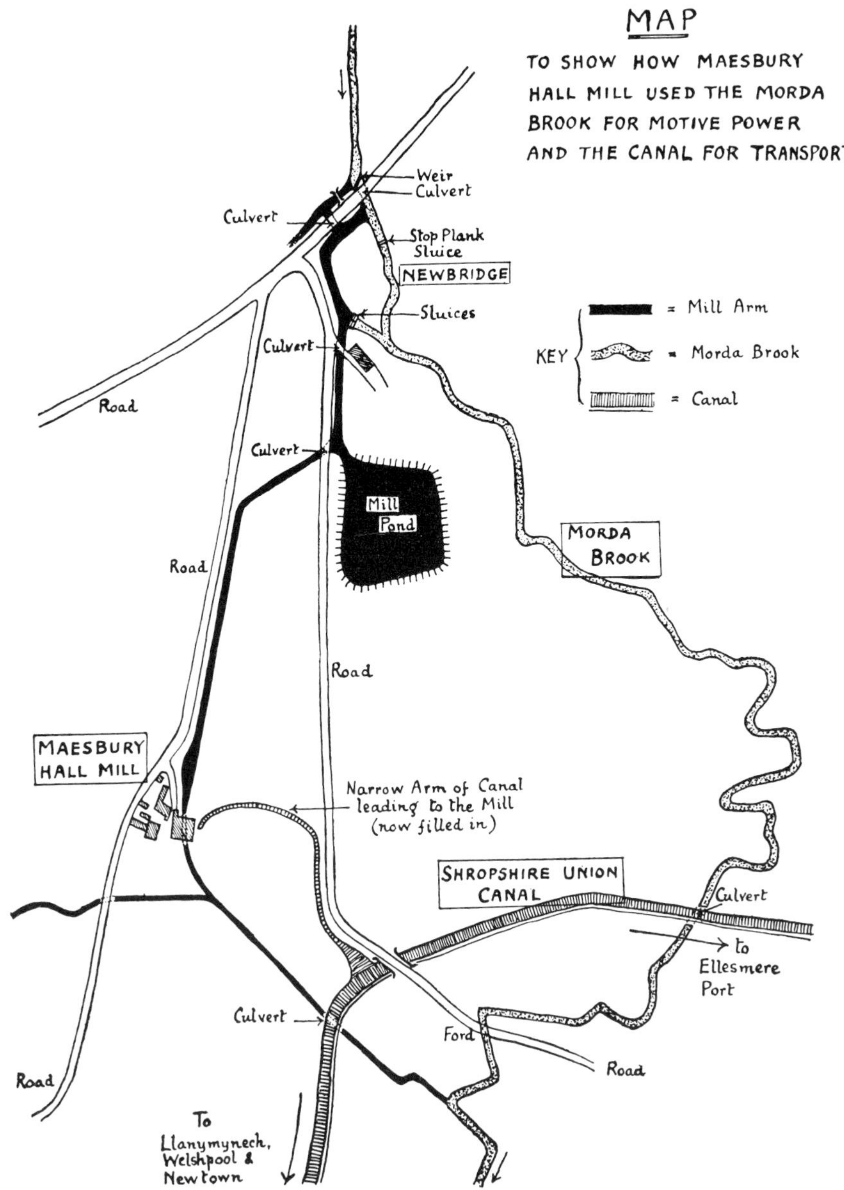

too damp for safe storage, and thus has to be dried and stored until such time as it can be milled.

Grain is obtained from a radius of some 40 miles around the Mill, but much grain is imported from Canada via Liverpool. Many years ago the Shropshire Union Canal provided the best means of transport between Maesbury and Liverpool, and Mr. Frank Peate was kind enough to give me the following information on this subject.

Until 1921 the Shropshire Union Railways and Canal Company (later the L.M.S.R.) did their own carrying in boats built and owned by them. After that date they ceased to do their own carrying, and the Peates bought a fleet of eleven narrow-boats from them, most of the boats being named after famous victories in British history, such as "Cressy," "Aboukir," etc.

These boats were family boats with a cabin at each end, in which lived the boatman, his wife and children. The "boat people" were a race apart, both in manners and dress, and were poorly educated, because, in spite of the fact that the children had to attend the local school if they stayed for a day or more in one locality, there was no continuity in their education.

Old warehouse and crane by the side of the canal, the surface of which is almost covered by flowering Frogbit. Part of the Navigation Inn is seen beyond the bridge.

Corn brought to the mill came from various sources. That from Canada was landed at Liverpool, where the procedure was as follows The grain was transferred in bulk to 100 ton barges, and these took it across the Mersey estuary to Ellesmere Port at the beginning of the old "Ellesmere Canal." Here it was discharged in bulk into the large warehouses built to store such cargoes until the narrow-boats came alongside to be loaded up. Two-cwt. sacks were then filled with the grain, and loaded into the boats until they contained their usual complement of twenty tons.

The long journey to Maesbury by canal then began. Across the Wirral Peninsula to Chester; along the old "Chester Canal" to Hurleston Junction near Nantwich; then by the old "Ellesmere Canal" again via Wrenbury, Whitchurch, Ellesmere, Welsh Frankton, Queen's Head and Aston Locks; and at last Maesbury was reached, after negotiating forty-one locks.

At the mill the sacks of grain were hoisted up from the boats, and emptied into the bulk stores. It took three days to get from Maesbury to Ellesmere Port, two days back, with one day for loading and another day for unloading.

When the Peates took over the boats they also took over their crews, paying them on tonnage, 2s. 10d. per ton. About twenty tons was carried per week by each boat, so that if nothing else was carried each boat would earn about £3 per week. However, it was usually arranged for empty boats to pick up other cargoes, such as road stone from Vron works, and deliver these where they were wanted. In this way earnings might reach over £5 per week.

By the year 1930 there were few people using this canal for commercial transport, practically the only people besides the Peates being the following coal merchants: Owen and Sons of Pant, Hydes of Welsh Frankton and Edwards of Maesbury Marsh. The coal came from the Black Park Colliery's Wharf near Chirk, and the Peates also obtained 20 tons per week for use in their Mill.

With so little revenue coming in by way of tolls, the canal authorities neglected the maintenance of the canal to such an extent that the canal began to silt up. The boats, when fully loaded, touched bottom in many places, making progress extremely slow and difficult, so that merchants had to lighten the loads carried until at last this method of transport became uneconomic. In 1932 the Peates stopped using boats and sold their fleet to the Leeds and Liverpool Canal Company and others.

During their more prosperous days the Shropshire Union Canal Company had built warehouses and depots at Newtown, Welshpool, Berriew, Maesbury Marsh, Ellesmere and Whitchurch, all solidly built and substantial buildings, of which that at Maesbury Marsh was rented by the Peates. Fly Boats (the express service of the canals) did the journey from Newtown to Ellesmere Port in two days, travelling day and night, and changing horses frequently. Stocks of grain, flour and many other goods were kept at the depots, and when the Shropshire Union Canal Company ceased carrying, the Peates sent boats to restock some of the depots occasionally. Other people kept stocks there also, and the former storekeepers were often kept on.

When the boats owned by the Peates were in need of repair they were sent to Welsh Frankton. Here there was a boat-building and repair dock and workshop just beyond the third lock. The boat builder was Mr. Beech. He repaired and painted the "Lesley," a second-hand boat; this and the "Margaret," built for the Peates by Mr. Nurser, of Braunton, were named after their children. He also altered the "Cressy," fitting it with a steam engine and later with a Ford car engine. Eventually the "Cressy" was acquired by Mr. L. T. C. Rolt, who travelled in her along most of the canals of England. An account of his journeyings, in which he was accompanied by his wife, are described in his delightful book, *Narrow Boat!.*

I asked Mr. Peate if he would use the Canal again if it were re-opened and kept in repair. He said that apart from the fact that this was not likely to happen owing to the lowering of certain road bridges where the water of the canal was piped through, and the piping of water for a distance of over a mile at Pant; there were good reasons for not using canal transport in his case.

The basic reason was one of "manpower." Too many men had to be employed on too many boats. If the canal had been made much wider and deeper so that at least 50 ton loads could have been taken, the position would have been different.

To-day the Peates employ bulk grain carriers which can carry a load of 13 tons of grain in bulk, and make two journeys to Liverpool and back on most days by road. Only two men are required and the carriers can be loaded up in a quarter of an hour, and unloaded in twenty minutes. On an average they bring 130 tons of grain each to the Mill every week.

To bring the same quantity of grain by canal used to require seven boats with their seven families, and, although each boat carried about 20 tons, it took a long time and the grain had to be filled into sacks at Ellesmere Port and these had to be emptied again on reaching the Mill. Further, the men were paid according to the tonnage brought, and, in addition there was the Canal Company's tolls to pay.

The lack of good roads was one of the major reasons for the construction of the canals, and they afforded a means of transport for any industries established along their banks. The Shropshire Union Canal played an important part in the growth and development of Maesbury Hall Mill, and without its aid the Mill would have remained an isolated country mill until its inevitable closure. It is rather ironic that the modern development of road transport has caused the closure of the canal and led to an increase in the prosperity of the Mill.

Shropshire's six "Inclined Planes" – the engineering wonders of the Canal Age

by W. HOWARD WILLIAMS

IN her series of articles, *Seen from the Towpath*, Lilian Hayward showed how intricate was the network of canals that once served Shropshire.

It is an undisputed fact that, in this county, at any rate, the invention of the inclined plane revolutionised this mode of transport.

Shropshire had been tardy about canal-building, owing, no doubt to the general unevenness of the terrain, and shortage of water for locking.

The Marquis of Stafford's canal, between Donnington Wood and Pave Lane, Newport, built about 1782, was the only Shropshire canal in existence when William Reynolds, the celebrated founder of the Ketley Ironworks, decided to build a canal between Oakengates and Ketley. The object of this canal was to speed up supplies of coal from the Oakengates district to the Ketley Ironworks.

Reynolds was faced with the problem of a

The Inclined Plane at Trench

73 feet fall in the level of the ground between Oakengates and Ketley, added to which was an insufficient supply of water for lockage.

Even if the water had been available, the number of locks required would have rendered the project uneconomical. Until this time, locks had been the only known method of conveying boats from a higher to a lower canal level.

The problem was solved by Reynolds in the ingenious invention which became known as the inclined plane. Telford, the great engineer, wrote in glowing terms about the device.

The method Reynolds adopted was to bring the canal forward to an abrupt part of the bank. Here he built a double lock into which ran a double set of rails which were laid on the face of the incline and which ran into the lower canal.

On each set of rails was an iron frame called a cradle, which was large enough to receive a 5-ton tub-boat—the type most generally used on the Shropshire canals. The wheels on the cradles were larger in front than at the rear, varying in diameter according to the ratio of the incline, so that the boats could be carried up and down in a practically horizontal position.

When a cradle carrying a boat entered the lock at the top of the incline, the lock-gate would be closed and the lock flooded until the boat was floated off the cradle into the canal basin.

Another boat would be brought on to the cradle and the water would be drawn off, the boat secured and with the lock-gate opened the boat would begin its descent to the lower level.

In its earliest days, before steam power was applied, a full boat descending would draw up a boat nearly equal to $\frac{3}{4}$ its weight. Later a steam-powered stationary engine supplied the power needed for drawing a fully laden boat up the incline.

The success of the Ketley inclined plane, which with the canal was built in 1788, was immediate, and to mark its completion a "token" was struck showing the imprint of the "Inclined Plane, Ketley, 1789."

The resource of Reynolds fired the imagination of other local industrialists, and in 1788 an Act of Parliament sanctioned the building of the Shropshire Canal between Donnington Wood and Coalport.

The building of this canal was a remarkable achievement when one considers the lie of the land it traversed.

Beginning at the Donnington Wood end of the Marquis of Stafford's canal, it almost immediately ascended an inclined plane 320 yards long to a height of 120 feet.

Passing through Wrockwardine Wood and

The Inclined Plane at Coalport.

the Snedshill Ironworks, it connected with Ketley Canal near the Greyhound Inn at Oakengates. Its course then lay across Hollinswood to Southall Bank.

The Aqueduct, near Dawley, carried a branch to the old workings at Brierley Hill, near Coalbrookdale, while the main canal went via the Windmill Farm incline, 600 yards long, falling 126 feet, continuing east of Madeley to the Haze above Coalport.

Here another incline was built, 350 yards in length, descending 207 feet to a short length of canal about ¾ of a mile in length which ran parallel with the river.

The gradient of this incline was about 1 in 3, and occasions have been recorded of the chain attached to the cradle breaking, causing the cradle and boat to rush violently down the incline and empty its contents (often 5 tons of pig-iron) into the river.

The Shropshire Canal was begun in 1788 and was completed in 1792, the cost of its construction being £47,000.

The ever-increasing cost of carrying coal to Shrewsbury led to the building of the Shrewsbury canal. Beginning at the point where the Shropshire canal began the ascent of the Donnington Wood incline, this canal passed near the old glass-house and Bullock's Flour Mill at Wrockwardine Wood to the Trench, where yet another inclined plane, 223 yards long, falling 73ft. 6in. was built.

From the Trench this canal proceeds via Eyton, Longdon and the Berwick Tunnel to Shrewsbury, a distance of 17½ miles. The locks on this canal are the smallest in England, measuring only 81 feet long by 6ft. 4in. wide.

The Trench inclined plane was the last of the Shropshire series to cease operating. In its latter years wheat was brought by canal to the Donnington Wood (Bullocks) Flour Mill, but the ever-increasing cost of maintenance, together with the closing down of the pumping engines which supplied the water (from the pits), brought an end to its usefulness in September, 1921.

Its last operator, William Jones, to whom the writer is indebted for the original of the illustration of the Trench incline, is still hale and hearty in his 81st year. He saw service on the incline from the early 1880's to its closing in 1921.

A sixth inclined plane was erected to serve the lime-stone quarries belonging to the Marquis of Stafford at Lilleshall.

It is unfortunate that these interesting and, at one time, revolutionary devices should have been allowed to disappear without more records of their former usefulness having been kept. With the exception of the Trench incline, it has been so far impossible to trace, with any degree of accuracy, the dates when the Shropshire inclined planes ceased to function.

If any reader can supply any concrete evidence as to when these inclines ended their careers, the Editor would no doubt be only too pleased to publish it.

The Canal System of East Shropshire and the Industries it served

by W. HOWARD WILLIAMS

WHEN Francis, fourth Duke of Bridgewater, conceived the idea of making a short Canal connecting the coal mines on his Worsley estate with Manchester for the purpose of supplying the latter with coal at a cheaper rate than had previously been possible, he little dreamed that his venture would earn him the title "Father of Inland Waterways." This ten mile long canal was completed in 1761, and so immediate was its success that many hundreds of canals were made in the ensuing years.

The first canal in Shropshire was that constructed by the Earl Gower and Company in 1768. There is little doubt that the Earl, who was the owner of much land in East Shropshire, had been inspired by the success of the Duke of Bridgewater's venture. The The Duke and the Earl were brothers-in-law, and their respective land agents were the brothers, John and Thomas Gilbert.

It had long been known that rich deposits of coal and iron-stone lay beneath the Earl Gower's Shropshire estate, which included the greater part of Donnington Wood, Wrockwardine Wood and Ketley. Lack of good roads or waterways had delayed the development of these industries.

In 1768, John Earl Gower took into partnership the brothers Gilbert, and this Company constructed what has been variously described as the Marquis of Stafford's or the Duke of Sutherland's Canal, but it was generally referred to as the Donnington Wood Canal.

The Earl Gower, it should be explained, was created first Marquis of Stafford in 1786. His eldest son, who became the second Marquis of Stafford in 1803, was created Duke of Sutherland in 1833.

Originally it was about five miles long, its Western extremity being at Donnington Wood. Passing South of Muxton village and almost under the shadow of Lilleshall Abbey, its Eastern terminus was near to the Newport—Wolverhampton turnpike at Pave Lane, two miles South of Newport. Here a wharf was constructed for the distribution of coal to Newport and the surrounding villages. Old maps reveal that lime was formerly burned in kilns close to the wharf. At a later date it was decided to link up the lime quarries and kilns belonging to the Earl Gower which had been established some years before in the vicinity of Pitchcroft and Lilleshall, with the canal.

A branch canal with several short arms was cut through the meadows North of the main canal. This branch was seventy feet below the level of the main canal, and after dividing at Willmer Bridge the Eastern arm passed through seven locks on its way to a stone pit at Pitchcroft. The Western arm ended at the North-Eastern end of Lilleshall village, near Limekiln Lane. Near to this terminus (which was called Colliers' End) were extensive lime quarries. A wharf was made here for distribution of coal to the village.

In the meadows between the East and West arms were the lime kilns. To service them a short cut was made from the Eastern arm, and a light railway ran out from the wharf at "Colliers' End" on the West arm.

To affect a connection between the main canal and the branch the latter ran into a short tunnel, whilst two shafts were sunk into the tunnel from the side of the upper canal. Coal was lowered in box-shaped containers by means of a hoist erected over the shafts into the boats in the tunnel. A container of coal descending one shaft would raise a more lightly loaded container of lime or limestone up the other shaft.

The lime would be sold at the various whavres on the canal for agricultural purposes, whilst the limestone would go to the Donnington Wood Furnaces which Richard Reynolds of Ketley had erected at the Western extremity of the canal, in 1772. In later years the limestone would undoubtedly be supplied to the furnaces at Wrockwardine Wood and the Lodge.

About 1790, the shafts and tunnel were abandoned and an inclined plane, 123 yards long was substituted.

When the original Lilleshall Partnership was formed in 1802 (later to become the Lilleshall Company), in which the Earl Gower was chief partner, the Lilleshall limeworks, the canal and all its appurtenances were leased to the new partnership.

When this company erected the famous plant of five furnaces at the Lodge, a short arm was constructed to serve that undertaking. These furnaces brought the company world-wide fame as here they produced the finest quality cold blast pig-iron ever made. The Lodge Arm was abandoned when the last furnace was "blown out" in 1888.

The beginning of the end of this canal came in 1878 or 1879, when the inclined plane and the branch canal were abandoned. It is safe to assume that this coincided with the closing of the Lilleshall Limeworks. Little use was made after this time of the section of the canal East of Muxton Bridge.

This was once the Lilleshall Inclined Plane with a drop of 70 feet.

About the year 1890, the Duke of Sutherland made his ong carriage drive between Lilleshall Hall and the Golden Gate. To accomplish this, it was necessary to destroy some hundreds of yards of canal, thus cutting off the section of the canal East of the drive to Pave Lane Wharf.

Boat-building and repair sheds were at the Old Yard, Donnington Wood.

By 1904, the remaining portion of the Donnington Wood Canal had become redundant. The Lilleshall Company were making increasing use of their mineral line from Muxton Bridge to Lubstree Wharf on the Humber Arm of the Newport Branch, with the object of saving time lost when the boats were passing the Trench Inclined Plane and the nine locks to Wappenshall.

The furnaces which Richard Reynolds had erected at Donnington Wood, having passed into the hands of Bishton's of Snedshill in 1796, were brought into the Lilleshall Partnership in 1802 by John Bishton, and thus became the Lilleshall Company's first furnaces. Though these works were eventually replaced by the Lodge Furnaces, the works were maintained as a foundry and engineering department until 1860, according to Randall.

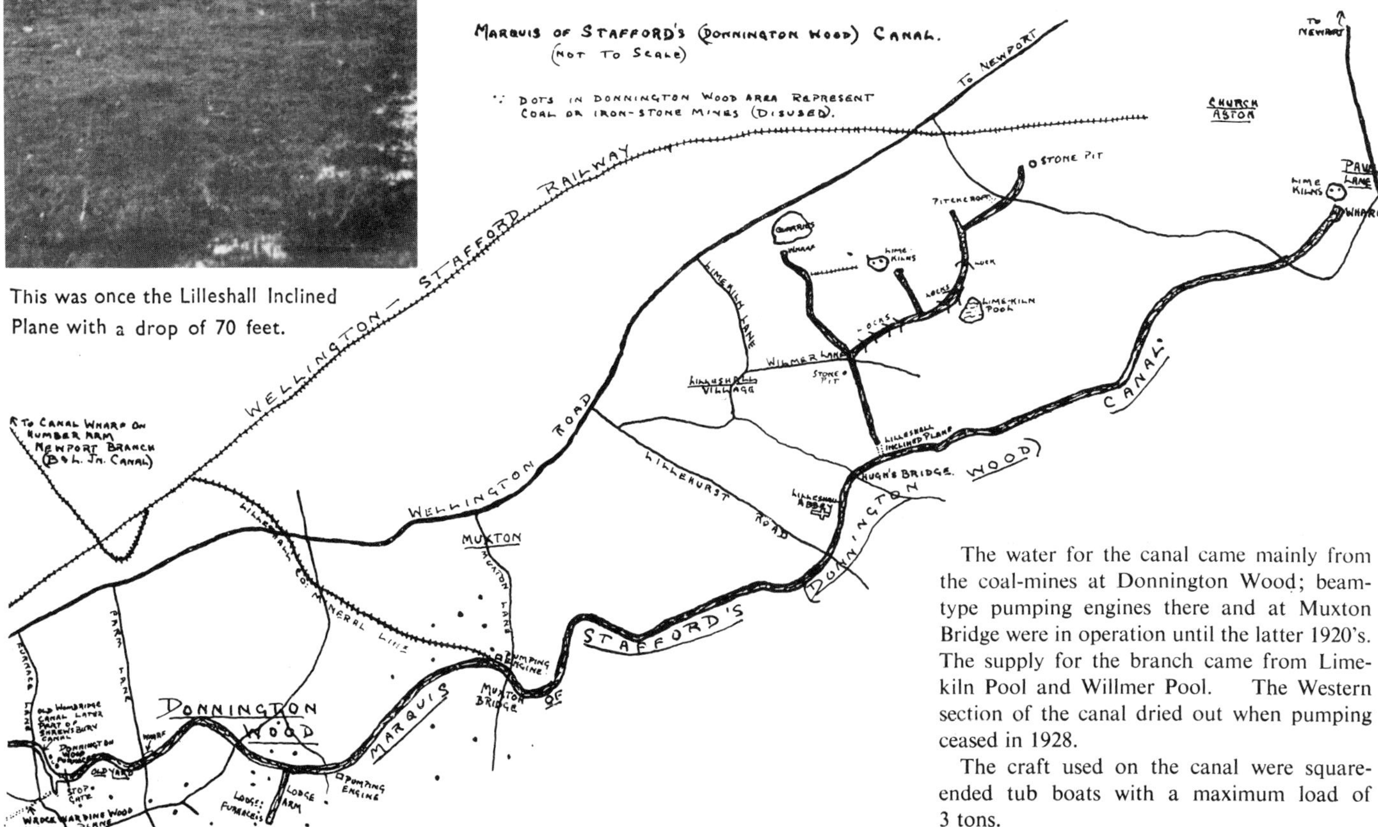

The water for the canal came mainly from the coal-mines at Donnington Wood; beam-type pumping engines there and at Muxton Bridge were in operation until the latter 1920's. The supply for the branch came from Lime-kiln Pool and Willmer Pool. The Western section of the canal dried out when pumping ceased in 1928.

The craft used on the canal were square-ended tub boats with a maximum load of 3 tons.

The Wombridge and Ketley Canals and the first British Inclined Plane

by W. HOWARD WILLIAMS

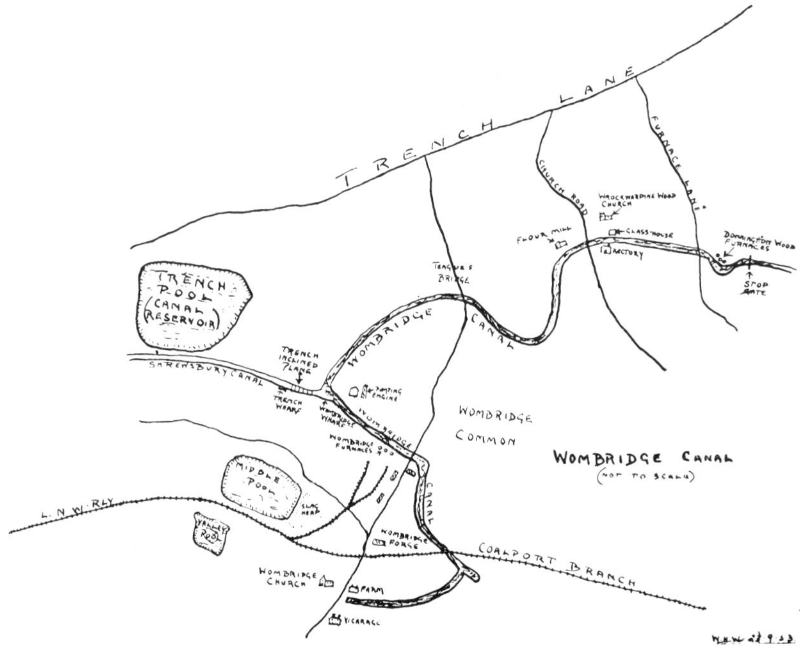

IN the previous article we saw that the Earl Gower and Co., were the canal pioneers of Shropshire. They were alone in the field for at least 19 years. By this time, 1787, Richard Reynolds, of Ketley, had handed over the management of his ironworks at Ketley and Donnington Wood to his sons, William and Joseph.

About 1787, William Reynolds discovered valuable desposits of coal and ironstone near the surface of the ground at Wombridge. Plimley (General View of the Agriculture of Shropshire, 1803), tells us that William Reynolds, "put into practice an idea he conceived some years before, of uncovering the strata of ironstone and coal which lay near the surface of the ground, so as to get the whole of the strata to a certain depth."

Lack of a good road between Wombridge and the Donnington Wood furnaces presented no problems to Reynolds. He constructed a canal about one-and-a-half miles long which began just south of Wombridge Church and ended in a junction with the Donnington Wood Canal. The Wombridge Canal, which cost only £1,640, was on a level throughout.

By a stroke of good fortune, the newly-formed Shrewsbury Canal Company in 1792 sought to purchase a little over a mile of this canal and incorporate it in the Shrewsbury Canal. Reynolds parted with it for £820—half its original cost. There is little doubt that by 1793 Reynolds had taken all the coal and ironstone he wanted from Wombridge and was pleased to find a purchaser for the canal.

In its time, this canal served three industrial undertakings on its banks. Close to where Wrockwardine Wood Church now stands, the Reynolds brothers erected a glass-house sometime before 1792. Most of the objects were made of dark green glass, such as door-stops jugs, rolling-pins and bottles for the French wine trade. The glass-house ceased to exist between the years 1835 and 1842.

Less than 300 yards from the site of the glass-house stands Donnington Wood Flour Mill, erected on the side of the canal in 1818 by John Bullock. The mill is still a very flourishing concern which, until a decade ago, was still owned by the family who founded it.

At Wombridge, James Foster erected two furnaces near the canal in 1818, and added a third at a latter date. In 1843 he closed the Wombridge Works and transferred the plant to Madeley Court. The Wombridge Works were opened again in 1854 by John Bennett for the Wombridge Iron Company and a forge was added. Until its closure about 1900 the firm was famous for the manufacture

Trench Lock, the highest lock on the navigation,

district owing to surface streams having been tapped at source during many years of mining. Even had the water been available for the nine or ten locks necessary to overcome the fall in contour, the venture would have been uneconomical.

The canal was begun in 1787 and was level to the south side of Ketley Hall. The 73 feet fall was overcome by the adoption of an inclined plane between the upper and lower canals. The idea was not a new one, although it was the first inclined plane on any British canal. Reynolds has been credited, erroneously, with the invention of this device.

The site of the inclined plane was in doubt for many years, but it is now certain that it was on the south side of Ketley Hall and falling westwards. What is not certain, however, is whether the lower canal went all the way to Ketley Ironworks or whether a

of wire and wire-rods. From 1904 the old Wombridge Canal was used only as far as Bullock's Mill, wheat supplies coming there from Ellesmere Port until the canal closed in 1921.

While Reynolds was making the Wombridge Canal, his mind was also occupied with other thoughts in the same vein. One of these was how best to improve the means of transporting coal and ironstone from the Oakengates district to his works at Ketley. The poorly metalled road between Oakengates and Shrewsbury, via Ketley, was constantly churned up by the passage of many coal carts, making them treacherous at all times and often impassable.

A canal was the obvious solution to the problem, but there were two snags that had to be overcome. The first was a fall in the contour between Oakengates and Ketley Works of 73 feet. The second was that there existed a general shortage of water in the

Old Wombridge Canal (incorporated in Shrewsbury Canal, 1793), Basin and Winding House. Note the two sets of winding gear, one for the incline and the other drawing the carriers over the sill. Drawn from an old photograph by J. F. Smith

light railway connected the works with the lower pond. The authorities, who hold the latter view, seem to be supported by a clause in the "Shropshire Canal Act" which refers to "a navigable canal and railway from Oakengates to Ketley Works."

Telford's article (date 1797) in Plimley's, gives the impression that the canal went to the works. The Ketley Tithe Map of 1840 does not reveal any railway, but a queer-shaped pool near the works has a slang of water, probably 50 yards long, reaching out eastwards in the general direction of Ketley Hall.. Incidently, this Tithe Map does not show the inclined plane. The fact that the map shows several pieces of the canal to be missing is evidence in support of the theory that the canal had been abandoned long before 1840, despite the fact that Bagshaw's Directory of 1851 states that "The commercial intercourse of Ketley is facilitated by the Shropshire Union Canal."

Dutens, a French engineer, found the inclined plane disused in 1818, and it is doubtful if it or the Ketley Canal were ever used again.

Telford, the great engineer of roads, canals and bridges, to whom we are deeply indebted for so much reliable information, tells us that the boats on the canal were of eight tons capacity.

They were similar in size to those later used on Shropshire and Shrewsbury Canals but were of greater depth.

Until the completion of the Shropshire Canal in 1792, the Ketley Canal had mainly a one-way traffic from Oakengates to Ketley The Shropshire Canal Act made provision for junction of the two canals and this was made just south of Snedshill Ironworks. The levels of the two canals were adjusted by a lock on the eastern end of the Ketley Canal.

The western end of a 20 yard long tunnel at Potter's Bank collapsed in 1897.

The Ketley-Coalbrookdale token, which was struck in 1789 to commemorate two feats of local engineering to gain world-wide fame, bears on its obverse side a replica of the Ketley inclined plane, with the Ironbridge on the reverse. The token was payable at Ketley and Coalbrookdale.

EAST SHROPSHIRE CANALS......3

The Shropshire Canal of 1788 was the first Public Enterprise of its kind

by W. HOWARD WILLIAMS

THE canals dealt with in the previous article were private undertakings. The Shropshire Canal, which was a much more venturesome enterprise, was financed by public subscription. It was a venturesome project because the contours of the country-side through which it was to be cut varied so greatly that no fewer than three large inclined planes had to be constructed, together with two long tunnels and one short one and a small aqueduct.

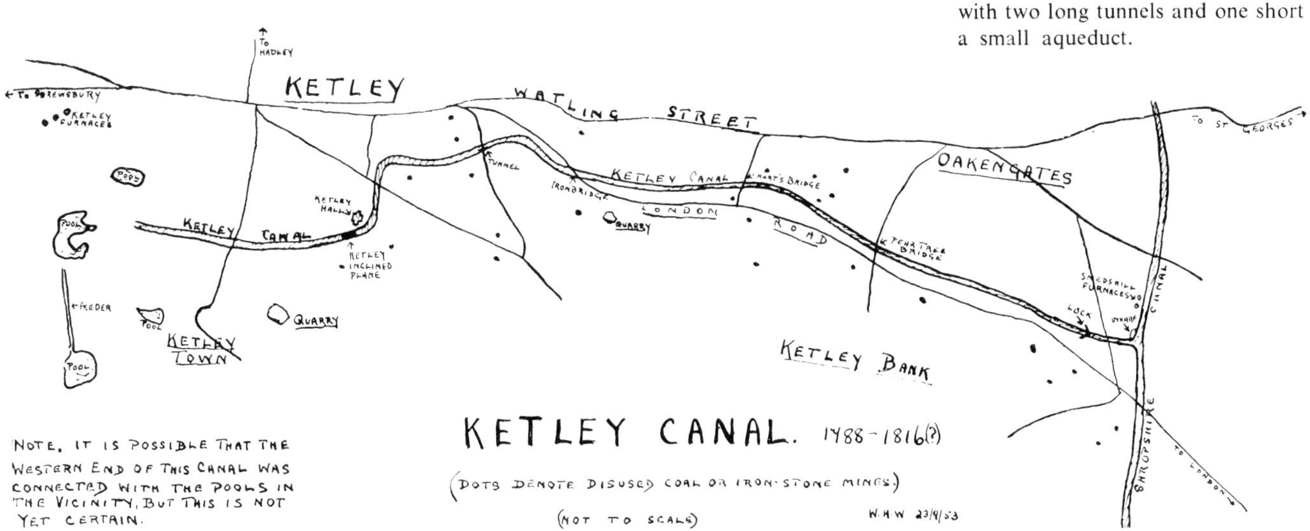

It is not at all surprising that the man who conceived the idea of making the canal, who surveyed it and did so much to bring the task to a successful conclusion was none other than William Reynolds, of Ketley. In a letter to his brother-in-law, dated January, 1788, he mentioned that he was making a canal from Oakengates to Ketley and had another in contemplation from Donnington Wood to the Severn.

In "A Dynasty of Ironfounders", Rastrick claims that this canal was a Coalbrookdale Company undertaking. A study of the "Shropshire Canal Act", 1788, Telford's article in Plymley's "Shropshire" and other authoritative works of reference, make it difficult to reconcile this claim.

In 1788, a subscription was entered into, and an Act of Parliament was obtained for the purpose of "making and maintaining of a navigable canal from Donnington Wood, to, or near a place called Southill Bank, and then by two several branches to communicate with the River Severn; one near Coalbrookdale and the other at Coalport".

There were to be 500 shares each of £125. The Act, which is a lengthy document, lists the owners of lands through which the canal would be cut, and gives the names of the 58 subscribers who formed the "Company of Proprietors of the Shropshire Canal Navigation". These include:— The Marquis of Stafford, The Earl of Shrewsbury, Lord Berwick, Sir Wm. Jerningham, Isaac Hawkins Browne, Robert Burton, William Cludde, Samuel Darby, Thomas Eyton, Thomas Gilbert, William Hallen, Egerton Leeke, William Pulteney, Richard and William Reynolds, Joseph, Elizabeth and Hannah Mary Rathbone, Hannah Rose, Cornelius Reynolds and John Wilkinson.

The canal began in a junction with the Donnington Wood canal, and almost immediately came the inclined plane 320 yards long, with a perpendicular rise of 120 feet. The canal then proceeded through

Site of the Wrockwardine Wood inclined plane, as it appears to-day.

Cockshutt, along the ridge of high ground above Wrockwardine Wood and Oakengates. After passing beneath the road between Oakengates and St. Georges, the canal passed through Snedshill Tunnel, 279 yards long. Thence it passed through Snedshill Ironworks, near which it was joined by the Ketley Canal. South of this junction, it passed under what is now the London — Holyhead Road, near to the Greyhound Inn by means of a short tunnel. About 300 yards south it passed near to Hollinswood Ironworks. Further south were the Dark Lane Furnaces of the Botfields on the west bank, while a little further on was Randlay Pool, one of the canal reservoirs. Nearby were the Stirchley Ironworks, also a Botfield Works. The Jerry Furnaces were on the opposite side of the canal, while about half-mile distant were Bishton's Langley Furnaces, which were served by a light railway. Nearer to Stirchley was another reservoir. The canal passed through the 281 yard long Stirchley Tunnel before it divided.

On emerging from the tunnel, the main line of the canal struck off across the meadows to Windmill Farm. Here a long incline, 600 yards long, was constructed. At the top of the incline were the boat-repair sheds. After passing the Windmill Incline, the canal passed under the road and into the Tweedale basin. From here light railways served the Madeley Court Furnaces, and the Halesfield and Kemberton Collieries. Water was brought to the canal here from those collieries.

The canal proceeded east of Madeley. Near the Queen Street Bridge, Messrs. Edge and Co., Chainmakers, once had a foundry, whilst Martin Randall's China Works once occupied a site near Hill's Lane Bridge. A light railway served Hill's Lane Colliery nearby. Further south, one can still see the gaunt remains of a bridge by which pit spoil was taken by a light railway from

Shawfield Colliery to Blist's Hill Brick-works.

Boat-building and repair sheds were situated at Blist's Hill. Here also were brick-and-tile works on opposite sides of the canal. The Blist's Hill Furnaces of the Madeley Wood Company, close by, were the last cold-blast furnaces to operate in the county. Randall records that, at a time of industrial strife in the middle of the last century, a body of strikers from other works in the district threatened to divert the canal into the works here, if the men did not stop work in sympathy with them.

Coal was brought to a wharf at Blist's Hill from the Meadow Pit, south of Madeley, by way of Baguley's Wind and the High Bridge over the Coalport Road.

The canal continued to the Hay, where the steepest of the inclined planes connected with a lower pound, 3/4 of a mile long, cut parallel with the River Severn, but above flood level, at Coalport. Between this pound and the river were erected warehouses, which were constructed so that goods could be conveniently transferred from canal boats to river craft and vice versa.

It has been estimated that goods including coal and iron, in excess of 100,000 tons passed through these warehouses annually in the peak of local industrial prosperity. It is an ironic thought that today Coalport is a forgotten inland port.

The branch canal, after leaving the main line, passed over a small stone aqueduct at a point where the Dawley—Madeley road passed through the small village now called Aqueduct. It then wound its way towards Dawley, passing and serving Dawley Castle Furnaces and Botany Bay Colliery, and several smaller collieries en route. Then on through Little Dawley to a point near to where Doseley church now stands. From here a light railway served Horsehay Furnaces. The canal then doubled back,

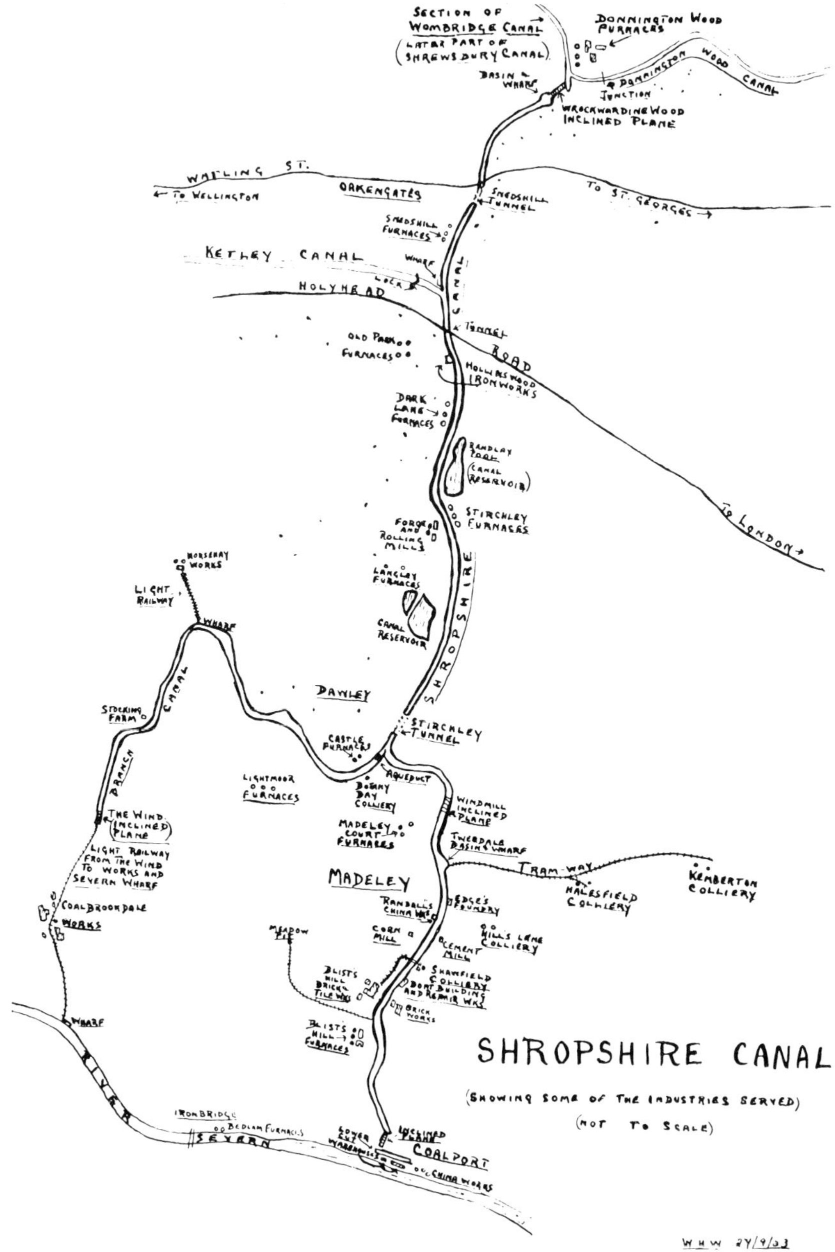

horse-shoe fashion, crossing beneath Holywell Lane and Green Lane to end at the top of the "Wind" at what was then known as Brierley Hill. The original plan was to take the canal over Lincoln Hill, near to the Rotunda, and to connect with the river near Styches Weir by means of an inclined plane, but this part of the plan was abandoned.

At the "Wind," a connection with the nearby Coalbrookdale Works was effected by twin shafts, 120 feet deep, sunk to a tunnel into which a light railway was run. The "Wind" operated on the same principle as a similar device at Lilleshall on the Donnington Wood Canal. This railway served the Dale Works and went on to the Loadcroft Wharf on the river near Dale End.

The canal and branch were completed in 1792.

In 1794, the Coalbrookdale Company abandoned the tunnel and shafts and constructed an inclined plane.

A map of Dawley Parva Manor, dated 1825, describes the section of the canal from the apex of the "horse-shoe" at Doseley to the top of the "Wind" as "dry canal". It had evidently been substituted by a light railway which is shown as running from the canal near Giggle Lane, through Lightmoor to Coalbrookdale.

remains of the bridge which carried the mineral railway from Shawfield Colliery to Blist's Hill brickworks;

the Aqueduct which carried the branch canal to Dawley.

In 1854 the canal was leased to the London and North-Western Railway, who planned to replace it with a railway. The necessary Act of Parliament to authorise this was obtained in 1857, and the Deed of Transfer is dated February, 1858. The Act authorised the closing of the section of the canal from the bottom of the Windmill incline to and including the Wrockwardine Wood incline, and the branch canal. The Coalport branch railway was opened in 1861.

The section of the canal from Tweedale basin to the river remained active until 1902, when the section from Blist's Hill Furnaces to the river was abandoned. The remaining section continued in use for the supply of fuel to Blist's Hill Furnaces from Kemberton and Halesfield Collieries, and shortly after the furnaces were blown out in 1912 the last part of this once very active waterway was abandoned.

The boats used were the square-ended tub boats carrying 5 tons, and they passed the incline at the rate of six up and six down per hour.

The actual cost of the Shropshire Canal and branch, which was about 10½ miles long, was £47,500.

According to "Fenn's Funds" (a primitive Stock Exchange Year Book) for 1837, the dividend paid on a £125 share was £8.

Voyage of 'The Tub,' the Last Boat through Preston Wide Hole

by R. C. EVANS

A fairly clear section of the canal, which for most of its length is choked by weeds.

I PULLED sharply at the toggle of the starting rope, the outboard motor for once co-operated immediately and began throbbing over quietly. The crew pushed off from the wharf and the boat slowly emerged from the half-gloom of the loading bay into the canal basin, warm in the July sunshine. On the banks three photographers stood, junior representatives of the West Midland press; they photographed our emergence from most angles, gingerly leaning over the banks to do so. Not content they ordered us to come out again, so round the basin, into the bay and once more into the sunlight.

We began at Wappenshall junction, one of the strangest-named and most quietly appealing of all British canal junctions. The quietness had intensified over the years, the canal being officially abandoned by Act of Parliament in 1944. The Shrewsbury Canal, leaving the main line of the Shropshire Union at Norbury, Staffordshire, descends seventeen neighbour locks into Shropshire; then locks its way through Newport to Wappenshall, where it is joined by the old Trench Tug-Boat Canal, two miles or so from Wellington. At Wappenshall the canal divides to form a basin branching before the basin, right under a bridge, left under a mill, the high roof of which formed the ten-yard tunnel which was our base.

The date of the last journey through the canal is unknown. Traffic ceased during the late twenties, although a maintenance engineer was supposed to have made a through voyage in 1936. In July, 1952, therefore, we were making minor history, the first mechanically-propelled boat in many years to travel the canal, and that on an abandoned waterway. The boat was an indefinitely-typed canvas and wood affair, flat wooden bottom, canvas sides supported by wooden struts, wooden combing about twenty feet long, and pointed bows and stern. It bore its name-plates, "The Tub," and port, Wappenshall Junction. It had a transom at one end supporting a second-hand four h.p. Britannia outboard. The boat had hung in a shed for years after an uncertain Home Guard career, and had been bought for £5. The crew were the 2nd Wellington Scouts, the destination as far as possible along the canal towards Norbury, eleven miles away.

In the Saturday afternoon sunshine the canal looked anything but abandoned. True there were rushes along each bank but a widish channel remained and there was a fair depth of water. The crew sat on a large chest filling the middle of the boat containing camping and cooking gear. We all felt, probably out of all proportion, that this was an adventure. The canal stretched out before the boat, infinitely more compelling than any road or railway, its course isolated, comparatively unknown, for canals are secretive affairs, and this one was officially dead.

At perhaps two miles an hour we approached the first bridge, a gradual narrowing of the view under the stone arch, holding the boat steady through the confined channel, then the echoing roar of the engine under the bridge and the emergence into the wide channel stretching away ahead.

Thls was bridge number twenty-eight from Norbury. The canal passed over Wappenshall Moor, low-lying farmland intersected with many streams. After about a mile we came to Preston Wide Hole, a turning point for narrow boats, reputed to be a pre-canal pool, bottomless, of course. Through the wide hole we approached Preston-upon-the-Weald Moors, a small village whose most interesting features are its name and a square-built hospital for widows and children, established in 1725.

At Preston Bridge a small crowd of children, sprinkled with adults, watched our coming, an awed silence breaking into obvious comments, "It's a boat," requests for a ride and general noise. After Preston the canal passed along a high embankment over the Moors, here we met a peculiarly reddish weed and masses of water-lilies. The boat ploughed through; we had a wire cage fitted over the propeller, which trailed weeds and sticks behind it and had to be cleared periodically.

On our left we passed the Lubstree Arm, almost completely weeded up, which once connected with a light railway at the "Humbers," a mile away. At the entrance was a lengthman's boat, a flat-bottomed broad punt, abandoned with the canal. This was lonely, or peaceful, country with few roads.

An occasional heron stood at the bank awaiting our approach, then flew further on to land and take up its stance again. At one time there were three, in line ahead, about fifty yards apart. Eventually each heron would tire of being moved on and fly away across the fields back to his original position.

Then we met the swans, undisputed masters of the water until that afternoon. They hissed furiously but moved on in front of the boat, eventually eleven of them. Suddenly they would fly into the air, once passing under and over a bridge, then, landing feet first, settle down again and swim furiously forward. Sooner or later they, like the herons, would fly back over our heads or turn and glide indignantly past us, though one stubborn bird swam ahead of the boat for over a mile.

Soon, too, we came upon the anglers, sitting heron-like spaced out along stretches of the canal. Some gave us good-humoured smiles, some soured looks. The latter obviously felt that we had no right to disturb the placid waters which were meant for fish, though there was no unanimity for we were asked to turn and stir up the fish which the afternoon sun had put to sleep. Occasionally a sleepy fisherman awoke sharply to pull his line out, bait long since nibbled from the hook.

There were three locks between us and Newport. We could not use them for the gates had been chained up for many years and the keys probably lost. The technique was to unload everything on to the bank, then the crew of six would manhandle the boat, carry it round the lock and load up again the other side. Locking through was an experience to come many times on canals and rivers still in commercial use; these Shrewsbury Canal locks, however, we could only gaze at with longing. The first was Longford Lock, named after a village, the second Polly's Lock, origin of name, and Polly, unknown. The canal Bradshaw, I think, calls Polly's, Edgmond Lock, but the Ordnance Survey map comes down in favour of Polly.

Between Longford and Polly's Lock the weeds became thicker until they stretched

The first lock on the Trench tub-boat canal at Wappenshall. The lock gate has been destroyed.

right across the canal. These "The Tub" charged and would emerge from the everglades trailing ten feet or so of water plants, brambles and, once, an old bicycle frame. Here we were met by a towpath squadron of five boys on bicycles who had followed us, it seemed, from Preston. They told us of a tree down across the canal ahead. It certainly looked formidable but moving gently through the weeds to the tree we pollarded the trunk with a "bushman's" saw and with the help of the dismounted cycle squadron pulled the wreckage clear.

We reached Newport Bottom lock at about eight o'clock, having taken four and a half hours to do seven miles. Slow, indeed, but Amazonian tributaries slow one down also. Before us the canal wound through Newport to the magnificent stretch, almost lock-free, certainly weed-free, to Norbury Bottom Lock, four miles away. But we had only the week-end and reluctantly pitched our tent in a clearing on the canal side. Supper followed, then a towpath ramble past the Newport locks, and bed.

Sunday morning was lovely, a mist came off the canal, the sun broke through and the temperature rose steadily. Bare-backed we cooked and ate breakfast, loaded up, and began the return journey. It might seem that retrogression would be monotonous but with canals the opposite is true. And this time, at least, the weeds trailed the other way. A family of moorhens kept us company for a time. Fishermen, looking as if they had been there all night, brewed early-morning cups of tea. We sometimes caught glimpses of fish in the water and pulled out two discarded small roach. We counted the number of streams, or "strines," natural and artificial, which drained the land on each side. The sun was very hot and horse-flies very fierce; killing them after they had pierced the skin, was only small satisfaction.

Twice the "pin" of the propeller-shaft was "sheared" on hidden rocks, causing the propeller to revolve wildly and horribly noisily. Each time the stern had to be lifted to the bank and a new pin quickly subsituted. This, the occasional brushing of the plug and the filling while moving of the petrol tank, was all the attention the engine needed. Slowly we chugged along—the canal is the most peaceful travel agent left in Britain, that is if you are not in a hurry, and to be in a hurry spoils travelling. We met the swans again, the ballet was repeated in reverse, the herons still kept watch. A kingfisher disappeared into an alder bush, a water-vole, nose showing, swam towards us,, leaving a diminutive wake. He sensed us and took refuge in a bank.

At Buttery Farm we moored and spent a leisurely two hours preparing, eating and digesting lunch. Then back across the moors, over the aqueduct above the Duke of Sutherland's drive, truly Roman in its straightness, unmetalled and unfenced. A crowd again saw us pass under Preston Bridge after an abortive attempt to explore the Lubstree Arm. Soon Wappenshall came in sight, the last bridge, and into the bay wharf once more. Beyond the bay the canal went on towards Shrewsbury, miles for another voyage; behind lay the last voyage. Now, the first bridge beyond Wappenshall has been levelled, the canal piped through, The way to the Preston Wide Hole and beyond is barred; we had been the last boat through.

Eyton's "Guillotine" to be sent to Transport Museum

by John W. H. Birch

QUITE recently British Waterways workmen have been busy dismantling the lock gate at Eyton-on-the-Weald-Moors, on the abandoned Shrewsbury branch of the Shropshire Union Canal. This gate is unique, the only remaining working example of a "guilotine" gate anywhere on the whole of the British canal system. As the name suggests, the movement of the gate is vertical, as compared with the door-type movement of the ordinary lock gate.

The Eyton guillotine gate, photographed before its removal for repairs and re-erection in London.

The importance of this gate is such that, after being renovated and put in perfect working order at Ellesmere, it is to be re-erected in the Waterways Museum, in Clapham, London.

The nature of the land made numerous locks necessary: many of these were of the guillotine type. Near to Leegomery there still stands a lock gate of this type, where the normal winch lifting gear is assisted by piling heavy rocks into a large iron-bound container.

The Eyton lock gate is not, however, the only unusual feature that the Shrewsbury branch canal—now only navigable for a few stretches, in a very shallow draught boat, and nearly waterless and in some places filled in for the last four miles into Shrewsbury — has to offer the canal enthusiast.

The first section of the canal to be built was that from Wappenshall Junction to Shrewsbury in the 1790's. This joined industrial Shropshire to Shrewsbury, the Shrewsbury Canal connecting at Wappenshall with the Trench "tub-boat" Canal, the outlet for a whole system of narrow canals.

Thomas Telford was the engineer responsible for the building of the canal. In characteristic manner he chose the straightest possible route. A major obstacle was the crossing of the River Tern, at Longden-on-Tern, and this he overcame by constructing the first-ever iron aqueduct, a small scale try-out for his revolutionary ideas which reached their fullest fruition in the great iron-capped aqueduct at Pontcysyllte. It is an amazing fact that, although it is weed-festooned, the Longden iron trough is still completely water-proof over 150 years after being constructed, whilst the framework was still strong enough to support the author, on his bicycle!

Nearer Shrewsbury, at Berwick, the canal passes through a third-of-a-mile long tunnel. This was one of the first tunnels to have a built-in- wooden pathway, to enable the barges to be hauled through by a gang of men, instead of being pushed and prodded along against the tunnel ceiling and sides. Although the one entrance is virtually blocked by a landslip, it is still possible to navigate the tunnel, although this is a somewhat dangerous undertaking and details and permission should be obtained from British Waterways.

Although the bulk of the cargo carried by the Shrewsbury canal was coal and iron, considerable quantities of barley (and probably hops) were brought to the canal-side maltings, at Ditherington. This building

Entrance to Berwick tunnel, partially blocked by a landslide. Navigation is "possible, though somewhat dangerous."

also is of considerable interest, for it was the first iron-framed construction to be erected anywhere in the World.

As has been mentioned, the canal which joined the Shrewsbury at Wappenshall was a tub-boat canal. The type of boat used on this canal, and the method of movement were for many years peculiar to the canals of Shropshire's "Black Country". The barges employed on these waterways were each about 14-feet long and 6-feet wide. They were towed in "trains", five being passed through the locks together.

But the most interesting, and still partially visible, feature of the Trench canal is at Trench itself, where the canal surmounts a considerable hill by means of an inclined plane. This remarkable example of man's ingenuity was, indeed, a Shropshire invention, the first one—now entirely disappeared, along with most of the canal it served—being built as part of the Lilleshall canal schemes. The idea of an inclined plane was to form a substitute for a chain of locks, the only other way to surmount a prolonged incline. In locks the barges stayed in the water all the time: on the inclined plane they were either fitted with wheels or dragged on to trucks, to be pulled up the rise and deposited in the water again. On the Trench canal the traction power—in later years a beam engine—was housed in a considerable brick structure on the hill top, which still stands—if in a somewhat dilapidated state —visible for several miles around. The gully up which the barges were hauled can still be traced, though much overgrown and denuded of the iron rails up which the containers ran.

During the 1830's the Shropshire Union Canal Company drove a canal between Wappenshall Junction and the main Shropshire Union Canal, at Norbury in Staffordshire, thus joining Shrewsbury and the Wellington areas with the main canal system. A century later only a few barges were making the trip to Shrewsbury: many of the tub-boat canals had been obliterated and built over for some years, and in 1944 the Shrewsbury branch of the Shropshire Union canal was official abandoned by Act of Parliament.

Decay has rapidly set in. The canal itself has become overgrown and unnavigable, with swans building their voluminous nests in 'mid-stream'. But this is perhaps not the greatest loss: for it must be acknowledged that modern means of transport are—at least in theory—faster and, therefore, more efficient than canals. The greatest loss is a link with the past. Still today on the side of the canal company warehouse at Longden-on-Tern a large, painted but now hardly readable sign proudly proclaims that "fly-boats" — the fastest means of transport on a canal and capable of travelling at up to 8 m.p.h.—ply daily from this point to Chester, Liverpool, Manchester, Birmingham and North and South Staffordshire. Nearer Shrewsbury, at Rodington, the romantic ideas of the 18-century creep into canal architecture, where the canal is spanned by a drawbridge, now not in working order and with its paint-work sadly neglected. Again, near Norbury a perfect Telford-designed lock-keeper's cottage now stands empty—in good condition, but subject like all houses to a steady deterioration if not tenanted.

Soon such hallmarks of the Canal Age as those mentioned above will crumble, and cease to exist. Much of the canal itself will probably be filled in to make field units of a size more suited to present day mechanized farming: during the past year this has indeed happened on the Lubstree branch of the canal, near Preston-on-the-Weald-Moors. The more important features of the canal will, like the Eyton lock, undoubtedly be preserved either in situ, or in a museum, by British Waterways themselves, or by the active Inland Waterways Association.

But the distinctive character which a canal gives to the landscape through which it passes, will almost certainly disappear entirely from many areas. The Shrewsbury branch of the Shropshire Union Canal, and the Trench tub-boat Canal—itself last of a whole system of similar canals—will be known to our descendants almost solely by particular and unique features — the Eyton Lock or the Longden aqueduct, the Berwick tunnel or the Trench inclined plane. Now, although not navigable, these canals can still be followed along their towpaths and the atmosphere of a past age can be effectively captured, merely by journeying along the banks and witnessing the various foibles and achievements of the Canal Age. Soon this opportunity will be lost.

The Shrewsbury Canal

By " KIWI "

IN a previous article in the Shropshire Magazine, I mentioned among waterways in Shropshire at present unusable the Shrewsbury Canal. This consists properly of the Newport Branch, the Shrewsbury Canal itself and the Trench Branch, but the whole can conveniently be referred to as the Shrewsbury Canal. It is the most truly Shropshire of all the canals, although it runs in Staffordshire for the first two or three miles of its course, not entering Shropshire until within about half a mile of Newport.

The Canal drops from its junction with the Main Line canal at Norbury by a magnificent flight of seventeen locks: some of the pounds between these locks are strikingly beautiful.

It then continues past the picturesque village of Forton and lovely Moss Pool, down Meretown Lock and Haycock's Lock to Newport. Here there is a spacious basin, but what could be a thriving wharf and marina is at present a desolate scene of mud, reed-beds and rubbish.

There is another lock at the Newport road bridge, while Tickethouse Lock and Polly's Lock follow at five hundred yard intervals. (Who, I wonder, was Polly? I like to think of her as a pretty lock-keeper's daughter, though her lock lies right out in the open country with now not a house nearby.) A mile later, Edgmond Lock brings the series to an end.

Soon after this, the canal begins to move out onto the Shropshire Weald Moors, a part of the county not widely known and with a charm of its own. There is a feeling of remoteness, which makes it hard to believe that the Oakengates conurbation lies only about two miles to the south. This part of the canal is raised upon an embankment, which is carried across Kynnersley Drive, (running between Kynnersely and Lilleshall) by a small, but exquisite aqueduct, well worth a visit.

After a picturesque section near the village of Preston upon the Weald Moors, our waterway moves on in gentle curves to Wappenshall Junction, where the Trench Branch leaves on the left, while the Shrewsbury Canal continues north-west. Here there are some wonderful old buildings in the true English canal tradition, not perhaps in the best of repair but still holding together.

A little way past the Junction is one of the characteristic Shropshire Union drawbridges still in reasonable condition and shortly afterwards the Canal descends by its last two locks near Eyton upon the Weald Moors. These locks, five hundred yards apart, are interesting on account of their guillotine bottom gates: these were hoisted with the aid of a counter-balance weight to allow boats to pass underneath.

The gate on Eyton Top Lock has disappeared, but that on the Bottom Lock, which stands less than half a mile east of the Wellington-Whitchurch road, is still in place, almost complete.

After the main road, the canal enters what is, for me, its most attractive section — here it winds through a beautiful, undulating plain, not so well wooded as to obscure the hills of the Wrekin massif on the one hand and Haughmond Hill on the other.

The River Tern is crossed at Longden by Telford's cast-iron aqueduct and the River Roden at Rodington by a sturdy and beautifully proportioned aqueduct of brick. The church at Withington stands right on the bank of the canal: what a picture this would make if only the canal were full!

Two and a half miles beyond Withington, the waterway plunges into the Berwick Tunnel, 970 yards long, 6 feet 5 inches high and nine feet one inch wide (measurements from "Canoeing", by William Bliss,

The beginning of the Shrewsbury Canal — it leaves the Main Line canal at Norbury and goes off under the bridge on the left.

To go through this tunnel in a canoe, by the light of a candle lantern, is an eerie experience. There is no towing path and in places water gushes down the sides; every now and again one looks up one of the wide vertical ventilation shafts and sees the sky as if from the bottom of a well.

Upon leaving the tunnel, the canal squeezes between the flank of Haughmond Hill and the left bank of the River Severn, which it now follows more or less closely, to run past the attractive village of Uffington into the Ditherington area of Shrewsbury. It originally ran to a terminus within a quarter of a mile of Shrewsbury railway station, but the last mile of its course has been lost in urban development.

Why should every effort be made to restore this canal? In the first place because we have one big headache of a national transport problem and it would be folly in the circumstances to ignore the potentialities of waterways which could help in any way towards the solution of that problem — a canal in being or restored to being is a valuable national asset; secondly, because the canal could provide an important supply of water for agricultural irrigation and other uses; thirdly because on aesthetic grounds we simply cannot afford to lose the pure functional beauty it could provide; and lastly because it could present a priceless amenity to anglers, pleasure-boaters, ramblers and the general public. Let our various local authorities (and also the Dawley Development Corporation) realise that this invaluable amenity runs past their doorsteps.

A correspondent in a Shropshire newspaper recently suggested that the canal be linked with the navigable portion of the River Severn at Shrewsbury. This would not be a difficult engineering problem and the potentialities for Shrewsbury would be enormous: the growing popularity of pleasure boating would cause the town to become a leading inland holiday resort; the probability of the revival of commercial traffic would be increased, especailly in view of the decreased coal-carrying facilities to be offered by British Railways; and a modest increase in trade would be experienced by people all along the route of the canal — especially by farmers, shopkeepers and hotel-owners.

The town of Newport would particularly benefit. It is also not beyond the bounds of possibility that the canal could in the future form an important link to a River Severn made re-navigable from Shrewsbury to the sea.

What would restoration entail? The main requirements are firstly a complete set of new gates for the twenty-five locks, the masonry of which is luckily in excellent condition. This is in contrast to the recently restored Stratford Canal, where thirty out of thirty-six locks required major repairs or re-building.

Owing to the removal of bridges, one

Longdon Aqueduct was an Experiment

ALTHOUGH the Shrewsbury canal has been out of use for many years this cast-iron aqueduct, which carries it over the River Tern at Longdon, remains as sound as the day it was made. It was designed in the closing years of the eighteenth century by Telford, the great engineer, and was really in the nature of an experiment. At the time it was constructed Telford was planning the much bigger aqueduct over the Dee valley at Pont Cysylte, near Llangollen, which is now regarded as his masterpiece. The idea of carrying the canal in an iron trough to save weight of masonry was suggested by Thomas Eyton, chairman of the Shrewsbury Canal Committee, but it was Telford who worked it out in detail. The trough, which was cast at Ketley, carries the canal on cast-iron pillars for 62 yards across the River Tern. The principle, having been put to the test at Longdon, was then applied with complete success to the Pont Cycylte aqueduct, which has a length of over a thousand feet and is 127 feet above water level.

The Main Line canal at Norbury — here is the life that any waterway should have.

main-road bridge and seven minor-road bridges would have to be built, as well as one minor road bridge raised to a height to permit full navigation.

Some farm crossings now with low fixed bridges would have to be equipped with lifting bridges as originally — one or two of the lesser minor roads once had drawbridges too, and there seems no reason why these should not be replaced.

At Pimley Manor, between Uffington and Ditherington, a stream crossing has been removed and the canal bed filled in for about eighty yards: this would have to be made good, an excellent exercise, one would imagine, for the Corps of Royal Engineers. A temporary basin would probably be required at Ditherington, which would complete the major works, exclusive of the proposed link with the Severn: the clearance of minor obstructions and of vegetation and the deepening of the small percentage of the channel that requires it would not present any serious difficulty.

At this point, I see hands raised in horror and the question asked: "But what about the cost?" I would hesitate to give a definite figure as I should probably be wildly wrong, the cost of materials would not be great because not a really large amount of materials — by modern standards — would be needed. The cost of labour would have to be minimised by the use, as in the Stratford Canal restoration, of voluntary labour, prison labour, the help of the Armed Services and similar expedients.

I would submit, however, that a nation which is prepared to invest £4,867,597 on six miles of motorway is likely to get at least as good a return for its money, proportionally, by investing a tiny fraction of that amount in the restoration of twenty-five miles of waterway. However vital motorways are to the country, and one would certainly not wish to dispute their importance, they will provide little in the way of peace and quiet and beauty and they will certainly never provide the peculiar efficiencies of waterborne transport.

As a final word on the subject of cost, may I quote from "The Canals of England" by Eric de Mare (The Architectural Press)? The quotation refers to Mr. J. F. Pownall's wonderful and far-sighted scheme for a Grand Contour Canal, but can also be applied to the present case:

"No doubt Treasury-bound officialdom will deride such splendid ideas and pose the eternal question: "Where is the money to come from?" Any reformer's answer must always be a definition of the word cost — "The total real cost of any project is not finally a question of money, since money should be merely a measuring system, but the amount of real wealth consumed by those concerned with that project during its execution. If that real wealth can be made available, the project can be accomplished. What is physically possible must be financially possible.'"

This canal was built by our fore-fathers with hand-tools and muscle-power. Are we to flinch from the infinitely lesser task of restoring it to use? There are signs that the waterways of this country are on the eve of a great resurgence. Is Shropshire to play no part?

Abandonment costs more than restoration

By "KIWI"

LAST July H.M. the Queen Mother performed the re-opening ceremony of the Southern Section of the Stratford-on-Avon Canal, which had been restored over the previous three years from a derelict state by the National Trust and the Inland Waterways Association working in co-operation, and using labour provided by volunteers, the Armed Services and prisoners. The Lower Avon Navigation Trust has re-opened navigation on the River Avon to above Evesham and there is reason to hope that navigation will eventually be restored all the way to Stratford. And now comes news that the British Waterways Board and the Staffordshire and Worcestershire Canal Society are to co-operate in restoring part of the Stourbridge Canal.

These are but a few, though possibly the most successful of the efforts so far, to rehabilitate Britain's waterways, and they have caused at least one Shropshire man to wonder if anything can or ever will be done for the waterways of his native county, including those running near its borders.

The principle Shropshire man-made waterway is the main line of the Shropshire Union Canal which runs along the eastern side of the county passing near Newport and Market Drayton. Originally know as the Birmingham Junction Canal, it was built by Thomas Telford in the eighteen-thirties, the last major canal to be built in Britain.

It is a beautiful waterway, fully functional, on which the Waterways Board is making steady improvements. It carries considerable pleasure boat traffic, and it does not seem too much to hope that, given a sane national transport policy, the commercial traffic, which it carried until quite recently, will return before too long.

The other usable Shropshire canal, though open now to pleasure traffic only, leaves the main canal near Nantwich and crosses north Shropshire to draw its water from the River Dee above Llangollen. It has a valid claim to be the most beautiful waterway in Britain, perhaps in the world. It was indeed officially abandoned in 1944 but, owing to the fact that it is now used to convey water to a purification plant near Nantwich, this wonderful waterway has been saved for pleasure boating and other amenities.

Let us now consider the waterways of present unusable, although the writer has spent many happy days, some of them not so long ago, enjoying their almost unbelievable peace and beauty as his canoe glided silently along them.

First let us take the Shrewsbury Branch, which leaves the main canal east of Newport and runs through that town and north of Wellington to near Shrewsbury.

Originally it consisted of the Shrewsbury Canal proper, which was dug from Ketley to Shrewsbury to convey coal from the Shropshire fields to the county town and to the River Severn it was linked to the old Shropshire "tub-boat" canals by the famous inclined plane at Ketley. Later the Newport branch was dug from the Shrewsbury Canal to link it to the main system.

The other, at present unusable waterway is the Montgomery Canal, which, running through the breath-takingly beautiful country of the Shropshire-Welsh border, is a close scenic second to the canal to Llangollen, which it leaves at Frankton, near Ellesmere and then passes through Llanymynech and Welshpool on its way to Newtown.

This canal suffered a small breach in 1936, which the then owners, the L.M.S. Railway, left unmended and Parliamentary agreement for its abondonment was obtained in 1944.

No survey of Shropshire's waterways would be complete without mention of the River Severn. Although now only navigable up to a few miles above Stourport, boats could at one time reach Pool Quay, near Welshpool, and to a considerably later date the river was navigable to Shrewsbury, as it still should be.

Can our Shropshire canals be saved as has the Stratford Canal?

An exact parallel cannot be drawn between the two. Our canals have been statutorily abandoned to navigation whereas by a fortunate chance the Stratford Canal had not: our canals have had several roads driven across them at water-level (an act of vandalism if ever there was one), whereas a proposal to do the same over the Stratford Canal was stopped in time.

On the other hand, our canals are not nearly so heavily locked and have not reached the advanced stage of dereliction of the Stratford Canal before its restoration. They would not on the whole pose such daunting difficulties as the restorers of the Stratford Canal were forced to face, and overcame so bravely.

The answer is, of course, that our canals can be saved if enough people want them to be saved and are prepared to do something about it, though every year that passes will tend to increase the difficulties. In the meantime the councillors of Newport, for instance, do well to be wary about the proposal of the Waterways Board to "de-water" the canal in that fine old town. Poor Newport! It has at present a "stinking ditch", when it could have a waterway which would be a thing of beauty and a real amenity, which would set off the picturesque character of the town as well as bringing a growing trade from pleasure-boaters, as has already happened at such places as Ellesmere and Market Drayton.

Our canals are part of the priceless heritage handed down to us by our forefathers, and we are in danger of losing them. The Inland Waterways Association advocate a policy of "Multifunctional Utilization" of the waterways as a means of saving them.

The following among others would be developed: the commercial transport function, as waterways provide the cheapest form of transport and could relieve the roads of much heavy and slow-moving traffic; the recreational function such as pleasure boating and angling; the function of beauty, so urgently needed in our largely urbanised community; the function of water supply (including agricultural irrigation), which is of growing importance in this country, and the parallel function of land drainage. Shropshire's canals rehabilitated could play their part in performing all these functions and more besides.

It is probably not generally known that it costs much more to abandon a canal than to restore it. The cost of restoring the Stratford Canal was £3,500 per mile. The British Transport Commission stated that the cost of abandonment would be about £10,000 per mile, while its elimination and complete filling in would have cost upwards of £60,000 per mile.

That is the case, put very briefly, for the Shropshire and border canals. Are they to continue as squalid leaky eyesores or are they to be restored to the functional beauty that was their's such a short time ago? The choice is ours.

Clee Hill Coal Was Too Dear

So the Leominster Canal Was Planned through South Salop

by M. Wight

THE enormous development of industry based on iron and coal in south Wales in the 18th century was largely due to English enterprise and to the development of canals, by which the products of the ironworks of the valleys could reach the ports. Among the pioneers was John Guest, who went in 1759 from Broseley to become manager of an iron furnace at Dowlais, which had been founded in the previous year. Thus began the long family connection with south Wales that still exists.

The earliest of the ironworks, set up about the middle of the 18th century, used the local iron ore and limestone, and burnt the wood of the Taff and Clydach valleys to make charcoal for their fuel. When Abraham Darby of Coalbrookdale had discovered how coal might be used instead, the south Wales ironmasters naturally turned to the coal that so fortunately was found close to the ironstone. A relic of this other connection with Shropshire can be seen in the name Coalbrookvale given to some ironworks in north Monmouthshire.

In 1782, John Guest invited his old neighbour Francis Homfray, who also owned ironworks at Broseley, and a forge on the river Stour at Stewponey near Stourbridge, to follow him into Wales; and he, with his sons Jeremiah and Samuel Homfray leased land near Merthyr Tydvil on which they build the Penydarren ironworks. Their works at Stewponey were served by the very succesful Staffordshire and Worcestershire canal, with its terminus at Stourport on the Severn. This had become a great inland port conveying goods from the Black Country, Manchester, the Potteries, Coalbrookdale and many other places to and from the great port of Bristol.

The aqueduct over the River Teme near Woofferton, blown up during the last war, makes an attractive picture. (S. B. Thomas).

As early as 1776 a small canal had been made near Merthyr, probably modelled on one built in 1769 for Lord Gower at Donnington in Shropshire. Between 1780 and 1790 there was a great increase in the number and size of the south Wales ironworks and collieries, and the need for better transport became more acute; at first iron and its products had been carried by pack horses or in waggons. So it is not surprising that canals were thought of. Thomas Guest and some of the Homfrays were among the chief shareholders in the Glamorgan canal, opened in 1794, although the road from Merthyr Tydvil down to the port of Cardiff had been turn-piked, with the Homfrays among its trustees.

The chief engineer of this important canal was Thomas Dadford, who had been resident engineer on the Staffordshire and Worcestershire canal and may well have been known to the Homfrays. Carriage by water was obviously quicker and cheaper than by road, although the construction of tramways began to speed things up at about the same period.

Then in 1829 Samuel Homfray made another step forward. He suggested the use of locomotive engines to replace horses on the tram road at the Sirhowy ironworks founded in 1794; and they began to run, at a maximum permitted speed of five miles an hour, before the end of the year.

Returning to Shropshire; the great success of the canal opened in 1772 from the Black Country to the Severn at Stourport produced several schemes to extend the benefits of inland navigation into Shropshire and Herefordshire.

Among several proposals put forward in 1777 was one for making a canal from the Severn at Bridgnorth to run down Corvedale to the Herefordshire border near Leintwardine and thence by way of Mortimer's Cross to Leominster on the river Lugg (which might be made navigable again as it had been early in the century) and so to connect with the Wye.

A survey was made of the country covered by this scheme by Robert Whitworth, one of the great Brindley's assistants, but his report of the probable cost was not encouraging. The matter was left in abeyance until 1789, when a meeting was held to consider a rather different proposal, for a canal to run from Leominster to Stourport by way of Woofferton and the Teme valley. As usual, the carriage of coal was the chief consideration. It was estimated that coal could be brought from Pensax by this canal to sell at 8/- a ton in Leominster; whereas Clee Hill coal, the only kind usually available was never less than 17/- a ton. It was agreed to apply for a Parliamentary bill to build such a canal, which would cross the Shropshire parishes of Ashford Carbonell, Burford, Boraston and Neen Sollars. In actual fact, it was never completed; only the first nineteen miles from Leominster was ever built, though this was when the canal mania was at its height; and a boat called "The Royal George," built at Stourport, was launched at Tenbury wharf. The *British Chronicle* of May 22, 1793, records this:—

"This being the first spectacle of the kind on the Leominster canal, a numerous concourse of people attended; and the launch took place amid the firing of cannon, flags, music playing and other demonstrations of joy."

In the same year the paper published a scheme for making a canal from the intended Montgomeryshire canal at Garthmyl to join the Leominster canal after passing through "Montgomery, Chirbury, Bishop's Castle, Hopesay, Onibury, Ludlow and Dirty Middleton." But this was rejected as too costly, and a modified scheme proposed, to run from Leominster to Lydham Heath. Neither ever materialised.

But the Leominster canal was opened between Woofferton and Marlbrook, the nearest point to Sir Walter Blount's collieries at Mamble. By 1796 the complete length of canal between Leominster and Marlbrook was open; fourteen barges with coal arrived at Leominster on the first day and it was sold at 15/- the ton.

Next year the price list gave "Best Cleehill Cornbrook coal 18/6 a ton, Lime coal, 15/- and Sir W. Blount's Ton Coal at 13/-."

Lack of funds was the chief reason for the failure to complete the scheme, though it was often discussed, until the coming of the railway between Hereford and Shrewsbury in 1858 put a final end to it. Some slight remains of the old canal can still be seen at Woofferton across a field opposite to the wireless station, where a cottage is named Lock Cottage and remains of an old lock are close to it. A weed-filled hollow between the station and the Salway Arms can be traced, and a mile away before Gosford Bridge are the remains of a three-arched aquaduct that carried the canal over the Teme; this was unfortunately broken down 'for security reasons' during the last war.

Index